Rock Engineering Design

Properties and Applications of Sound Level

Rock Engineering Design

Properties and Applications of Sound Level

Harsha Vardhan • Rajesh Kumar Bayar

CRC Press
Taylor & Francis Group
Boca Raton London New York

CRC Press is an imprint of the
Taylor & Francis Group, an **informa** business

CRC Press
Taylor & Francis Group
6000 Broken Sound Parkway NW, Suite 300
Boca Raton, FL 33487-2742

Printed on acid-free paper
Version Date: 20130722

International Standard Book Number-13: 978-1-4665-8295-8 (Hardback)

Library of Congress Cataloging-in-Publication Data

Vardhan, Harsha.
Rock engineering design : properties and applications of sound level / authors, Harsha Vardhan, Rajesh Kumar Bayar.
pages cm
Includes bibliographical references and index.
ISBN 978-1-4665-8295-8 (hardcover : alk. paper)
1. Rock mechanics. 2. Rock noise. 3. Sound--Reverberation. I. Bayar, Rajesh Kumar. II. Title.

TA706.V37 2014
624.1'832--dc23 2013026194

Visit the Taylor & Francis Web site at
http://www.taylorandfrancis.com

and the CRC Press Web site at
http://www.crcpress.com

The authors dedicate this book to their parents and family members

Contents

Preface

Engineering design of rocks is a complex process, requiring knowledge of their various physico-mechanical properties. There are established standard techniques for determining rock properties, which have been used around the world for quite some time.

The application of sound level in determining the fault and its diagnosis in the mechanical industries is very well known to scientists and engineers. However, its application in rock engineering design is not very common. Only a few studies can be traced pertaining to the application of sound level in rock engineering design. It is our intention to present to engineering students, scientists, and engineers the basic concepts and principle on which sound level can be used in solving rock engineering design problems in the clearest and simplest form possible. A major objective of this book is to help students to develop interest in using sound level as a tool in rock design applications.

This book has been developed out of many years of research experience gained by the authors. There are numerous books on the subject of rock mechanics and design. However, to our knowledge, a book specifically in the area of rock engineering design using sound levels is not available on the market today. This was the main intention in bringing out this book. The theory presented is as simple as possible to make the subject interesting to readers. The book is equally useful for readers with backgrounds in both mining and civil engineering disciplines.

Regarding its organization, the book consists of 10 chapters, well arranged in a coherent manner. All the chapters are presented in a logical order, starting with Chapter 1, which provides a general introduction to noise, its effects, and standards. Chapter 2 discusses the application of noise monitoring for mining equipments, whereas Chapter 4 discusses the application of acoustic emission techniques in geotechnical fields. The equipments for drilling, measurement of sound, and physico-mechanical properties of rocks are discussed in Chapter 5. The process involved in the measurement of rock properties and sound level is discussed in Chapter 6. The statistical values of the rock properties and sound level produced during drilling of different rocks are summarized in tables and figures in this chapter. Chapter 7 explains developed regression models, whereas procedure and the results of developed artificial neural network models are presented in Chapter 8. A case study carried out during this work is discussed in Chapter 9. Finally, a summary and suggestions for further work in this direction are given in Chapter 10.

This book focuses on the indirect method with emphasis on the development of numerical models in rock engineering design using sound levels.

This does not mean that the direct methods are not important. In practice, a project should always include some types of laboratory or in situ tests, and the indirect methods can only be used to supplement the direct methods.

We are grateful to various authors and publishers who permitted to reproduce their illustrations. We are equally grateful to the reviewer of the manuscript of this book who made extremely valuable suggestions and has thus contributed in enhancing the standard of the book. Further, we gratefully acknowledge the support and the cooperation of the staff of Taylor and Francis who were responsive to our wishes and helped to create the present layout of the books. We shall feel amply rewarded if the book proves helpful in the development of genuine research studies.

Harsha Vardhan and Rajesh Kumar Bayar

The Authors

Dr. Harsha Vardhan completed his B. Tech in mining engineering from the National Institute of Technology Karnataka at Surathkal and a Ph.D. from the Indian School of Mines, Dhanbad. Before joining the faculty at NITK Surathkal, he worked with Mysore Minerals Ltd. (a Government of Karnataka undertaking) for around 2 years. At present he is working as Associate Professor, Department of Mining Engineering, National Institute of Technology Karnataka at Surathkal and Mangalore. He is a fellow of Disaster Advances, and life member of the Institution of Engineers, India, and a member of Institution of Public Health Engineers, India; Mining, Geological and Metallurgical Institute of India; Acoustical Society of India; Mining Engineers Association of India; and Indian Society for Technical Education. Over a period of time, he has developed expertise in the area of mine environment, underground ventilation, and noise control of mining equipment. He has developed special interest in the application of acoustics in rock mechanics.

Dr. Vardhan is actively involved in research and consultancy, and he has to his credit more than 86 research papers in international and national journals and conferences. He is on the editorial board and reviewer of several professional journals in the area of his expertise. He has received the Award for Best Research Publication from the Government of Karnataka; Engineering Gold Medal and D. N. Thakur Award from the Mining, Geological and Metallurgical Institute of India; and Certificate of Merit from the Institution of Engineers, India. He was also honored with the Career Award for Young Teachers from the All-India Council for Technical Education, New Delhi, and Young Scientist Award from the Ministry of Science and Technology, Government of India.

His work has taken him on various assignments to the United States, France, Thailand, Belgium, and Indonesia. He is also involved in providing technical assistance to various government and statutory bodies, which has drawn national attention.

Dr. Rajesh Kumar Bayar completed his Bachelor of Engineering degree in mechanical engineering from Mysore University, his Master of Technology in advanced manufacturing engineering, and a Ph.D. from the National Institute of Technology Karnataka at Surathkal. Before joining the Technical Institute of Ministry of Manpower, Sultanate of Oman, he worked with the N.M.A.M Institute of Technology, Nitte, for around 10 years. At present he is working on the faculty of the Department of Engineering, Mechanical Section, Higher College of Technology, Muscat, Sultanate of Oman. He is a life member of the Indian Society for Technical Education. Over a period of time, Dr. Bayar has developed expertise in the area of CAD/CAM, finite element analysis, artificial neural network, and rock mechanics. He is actively involved in research and consultancy.

1

About Noise in General

1.1 Genesis of Noise, Its Propagation, and Unit

Sound is generated due to surface vibrations or due to turbulence of an air stream, which sets up rapid pressure variation in the surrounding air. A sound wave is a longitudinal vibration of a conducting medium such as air or water. Sound waves can be represented as sinusoidal patterns with given amplitudes and frequencies. The intensity of sound, which is related to its loudness, is represented by the amplitude of the sound wave. Sound waves require a medium to travel. The medium must have mass and elasticity, and this is the reason why sound does not travel through vacuum. When sound is produced, a transfer of energy from the source to the surrounding molecules takes place. The velocity of propagation of disturbance in a medium depends on its density and elasticity and is given by the equation

$$C = K\,(E/\rho)^{1/2}$$

where E = Modulus of elasticity of the medium, N/m^2

ρ = Density of the medium, kg/m^3

K = Constant

For air under normal conditions, the velocity of propagation of sound is approximately 344 m/s.

The rate at which this disturbance occurs is expressed in cycles per second or hertz (Hz), which is the international unit of frequency. A higher-pitched sound is one that is higher in frequency. The normal human ear is sensitive to frequencies between about 20 Hz and 20 kHz.

For the purpose of accommodating the large range of sensitivity of the human ear, where the greatest intensity that can be tolerated at the threshold of pain is 10^{13} times that corresponding to the threshold limit of hearing (20 μPa), a logarithmic scale is used, and the unit of measurement is the bel (B). It is defined as

$$1\ \text{Bel} = \log_{10}\,(P/P_r)^2$$

where P = measured root mean square (rms) pressure, Pa

P_r = reference root mean square (rms) pressure, Pa

The lower limit of the threshold of hearing (20 μPa) is taken as the reference pressure (P_r). The entire auditory range is covered in 13 steps, 10^0 to 10^{13}. One bel signifies a 10-fold increase in intensity, two bels signify a 100-fold increase in intensity, and three Bels signify a 1000-fold increase in intensity, etc. For practical use, bels are too big, and therefore the scale is divided into tenths, the units being known as *decibels*. It is defined as

$$1 \text{ Decibel (dB)} = 10 \log_{10} (P/P_r)^2$$

This scale is appropriate to human hearing system, as the human ear also responds in a logarithmic way.

Before measuring the sound level, it is processed through a *weighting network* to simulate the human auditory system. Various research groups have worked on weighting networks and termed them as A-weighting, B-weighting, C-weighting, and flat-weighting. Nowadays, the A-weighting network is commonly used by regulatory bodies. If the measured sound pressure level is X dB using an A-weighting network, then it is expressed as "The A-weighted sound level is X dB" or simply X dB(A).

1.2 Common Terminology

Exchange rate: Defined as "the change in sound level corresponding to a doubling or halving of the duration of a sound level while a constant percentage of criterion exposure is maintained."

Average sound level: The logarithmic average of the sound during measurement duration (specific time period), using the chosen exchange rate factor. Exposure to this sound level over the period would result in the same noise dose as the actual (unsteady) sound levels.

Noise dose: The percentage of time a person is exposed to noise that is potentially damaging to hearing. Zero represents no exposure and 100 or more represents complete exposure. It is calculated by dividing the actual time of exposure by the allowed time of exposure. The allowed time of exposure is determined by the criterion duration and by the sound level (the higher the level, the shorter the allowed time). The sound levels must be measured with A-frequency weighting and slow exponential time weighting.

Energy equivalent sound level (L_{eq}): The equivalent steady sound level that, in a stated period of time, would deliver the same amount of sound energy as the fluctuating or time varying sound during the same period. It can be calculated using the equation:

$$L_{eq} = 10 \log \Sigma f + 90$$

Where,

f = Fractional exposure factor = 0.125 t antilog [0.1 (L – 90)]

t = Time interval in hours

L = A-weighted steady sound level in dB for time interval "t"

Threshold sound level: The A-weighted sound level below which, the sound produces little or no noise dose accumulation and may be disregarded. It is used for hearing damage risk assessment.

1.3 Impact of Noise

Worldwide, increasingly stringent regulations coming into force are limiting the exposure of workers to industrial noise. As a result, industrial noise is gaining importance not only to employers but also to state and central government officials, trade unions, occupational hygienists, physicians, and insuring companies.

A number of machines and operations in mines produce high sound levels, which may have a significant effect on worker/operator health. Apart from health issues for personnel, noise is also known to cause significant economic loss to an organization. Daniel et al. (1998) reported that, in the state of Washington alone, the annual disability settlements for hearing-related problems approached nearly 22.8 million dollars.

The ill effects of noise can broadly be categorized under three headings: interference with speech communication, auditory, and nonauditory effects.

1.3.1 Interference with Speech Communication

Interference with speech communication is one of the most serious consequences of noise. ReVelle and ReVelle (1974) estimated some 30 to 40 million Americans feel speech interference after being exposed to certain sound levels.

The A-weighted sound level from 50 to 60 dB is known to affect normal conversation. If this level goes beyond 75 dB, then speech interference occurs (Pandey, 1978a; Pal and Saxena, 1999). Speech can be perceived most effectively, particularly as distinct from noise, when all frequencies from 300 to 6000 Hz are adequately presented to the listener (Stanley, 1969). But in a real situation it may be difficult as there is considerable variation in sound energy over the entire audible range of frequencies (Harris, 1979). This is particularly true for mining equipment with which considerable variations in sound levels are observed over the audible frequency range (Margaret, 1969).

A survey carried out in India regarding noise pollution due to open-pit coal mining activity indicated that around 47.3% mine workers and 48.9% officers feel speech interference during working hours (Dubey and Nath, 2000).

1.3.2 Auditory Effects

Auditory effects are the effects of noise on hearing systems. Wright (1964) indicated the first noted report on the effects of noise by Ramazzini in 1700, in *De Morbis Artificium Diatriba* that workers who hammer copper gradually became deaf. Labenz et al. (1967) measured the hearing of 66 operators of earth moving equipments. The results indicate considerable hearing loss among operators of all age groups as compared to those not exposed to it. Williams and Ross (1968) found that the effects of noise on hearing vary with the spectrum and intensity of the noise, the duration of exposure, the interval between exposures, the susceptibility of the individual, the presence or absence of other ear complaints, and whether the noise is intermittent or continuous.

Noise is known to cause both temporary and permanent threshold shifts. Powell (1956) indicated that any loss of hearing due to noise, which persists even after a period of 6 months, must be regarded as permanent. Glorig (1958) and Pandey (1978b) found that the loss of hearing sensitivity that results from exposure to continuous broadband noise, is first characterized by a decrease in sensitivity in the frequencies around 4000 Hz.

All people are not equally sensitive to noise (Walker, 1963). Aljoe et al. (1985) found that at the age of 60, over 70% of the miners had a hearing loss greater than 25 dB, and about 28% had a hearing loss greater than 40 dB.

The range of frequencies that can be heard by human beings is somewhat variable. Young people with normal hearing may hear all tones from 20 Hz to 20 kHz. All frequencies are not heard equally well at very low sound pressure levels (Davis, 1947). However, at a high sound pressure level, for example, 100 dB, all tones may be heard equally well. Most individuals develop an increase in hearing levels as a result of aging (Glorig, 1958). It is usually very difficult to distinguish increased hearing levels resulting from nonnoise factors, that is, disease or injury, and from those caused by exposure to noise. However, sufficient evidence exists to indicate that exposure over a period of years to noise whose sound pressure level exceeds 85 dB may cause both temporary and permanent elevations of threshold hearing levels (Rudmore, 1958). There is also evidence to indicate that a small number of individuals are highly susceptible to noise (Archibald, 1964; Barker and Gill, 1972; Gerald, 1975; Khuntia and Mishra, 1994; Singh et al., 1997).

1.3.3 Nonauditory Effects

Nonauditory effects are the effects of noise other than hearing loss. It has been reported by several researchers that exposure to high sound levels lead to increased heartbeat, indigestion, tension, anxiety, anger, and emotional imbalances (Franken, 1974; Miller, 1974; Broner, 1978; Krichagin, 1978; Peter and Bochum, 1978; Roychowdhury and Dhar, 1982; Das and Deshbhratar, 1982; Park and Park, 1983; Mitra, 1987; Job, 1988; Tiwari, 1992; Tripathy and Patnaik, 1994; Mukhopadhyay and Dey, 1999).

Hetu et al. (1988) reported anxiety, stress, isolation in groups, and negative self-image due to hearing loss. A similar study on industrial workers by Hetu et al. (1994) indicated "fear of social isolation" as the main reason of not reporting noise-induced hearing loss.

Leigh and Miller (1998) found that hearing loss accounted for more lost days than any other occupational illness. Singh et al. (1999) described a model showing the connections between noise reactions and health effects. Suter (2002) described a study conducted by Hallberg who interviewed wives whose husbands suffered noise induced hearing loss. A summary of the wives response indicated that the husbands' hearing loss was often the cause of misunderstandings and irritation within the family.

Noise not only affects workers' health but also reduces productivity (Neely, 1967; Hockey, 1972; Pujara, 1998; Bhatia, 2002). According to Niebel and Freivalds (1999), intermittent broadband noise can lead to decrease in productivity and increase in employee fatigue due to annoyance and distraction.

1.4 Statutory Provisions Pertaining to Noise Exposure in India and Other Countries

Assessment of the degree of noise harmful to a workplace is done by comparing the values measured at workplaces to the permissible threshold limit values (TLV) adopted by different countries.

National standards adopted by different countries lay down maximum permissible sound levels, depending on the kind of work being done and the length of exposure, with due consideration to two basic criteria, viz. (a) the threshold at which effects become noticeable (short-term action) and (b) protection of the health and working capacity of the workers, assuming lengthy systematic exposures. There are further two more criteria as well: (c) the optimum condition for output and quality of work and (d) a reasonable degree of comfort for the worker.

1.4.1 Noise Standard in India

In India, for general industries, ISO (International Standard Organization) code of practice is followed in terms of noise. ISO recommends a limit of 90 dB for the 8-hour average A-weighted sound level and halving the duration for every 3 dB increase in sound level (3 dB exchange rate) on the basis of equal energy criteria as given in Table 1.1.

To regulate and control noise pollution for mining occupations in India, the Directorate General of Mines Safety (DGMS) in its circular No. DGMS Cir.Tech./18 of 1975 and DGMS Cir.Tech./5 of 1990 suggests a warning limit of 85 dB and a danger limit of 90 dB for the 8-hour average A-weighted

TABLE 1.1

ISO Noise Standard

A-weighted sound level (dB)	87	90	93	96	99	102	105	108	111	114
Exposure time (hours)	16	8	4	2	1	1/2	1/4	2/15	1/15	1/30

TABLE 1.2

ACGIH Noise Standard

A-weighted sound level (dB)	80	85	90	95	100	105	110	115[a]
Exposure time (hours/day)	16	8	4	2	1	1/2	1/4	1/8

Note: Exposure to pulse or impact noise, should not exceed 140 dB (peak acoustic pressure).

[a] Exposure to continuous or intermittent A-weighted sound level louder than 115 dB is not permitted.

sound level. DGMS also recommends A-weighted sound level of 115 dB at and above which the unprotected ear may run a risk of hearing impairment and therefore appropriate ear protective devices should be used, and 140 dB where no worker should enter even with ear protection.

1.4.2 Noise Standard in United States of America

In 1972, the National Institute for Occupational Safety and Health (NIOSH) had set the occupational exposure limit to noise as:

The present limit for the 8-hour average A-weighted sound level is 90 dB, but 85 dB is urged and must henceforth be observed by all new plants. It also states that no worker should be exposed to A-weighted sound level in excess of 115 dB at any time.

The American Conference of Governmental Industrial Hygienists (ACGIH) proposed the limit for continuous or intermittent noise which is given in Table 1.2.

Occupational Safety and Health Administration (OSHA) has given maximum permissible exposure levels under Federal Code of Regulations 1972, which is given in Table 1.3.

When daily exposure to noise is made up of two or more periods of exposure to sound of different intensities, the sums of the following fractions

$$C_1/T_1 + C_2/T_2 + \ldots C_n/T_n \text{ should not be more than unity}$$

where

C_i = time of exposure for particular sound levels

T_i = permissible exposure time at that sound levels

TABLE 1.3

OSHA Noise Standard

A-weighted sound level (dB)	90	92	95	97	100	102	105	110	115
Exposure time (hours/day)	8	6	4	3	2	1½	1	1/2	1/4 or less

TABLE 1.4

Noise Standards in Argentina

A-weighted sound level (dB)	91–95	96–100	101–105	106–110	111–115
Exposure time allowed (hours/day)	6	4	2	1	1/2

1.4.3 United Kingdom

In April 1972, the Department of Employment and Productivity issued a code of practice for reduction of exposure to noise and laid down the limits of exposure. For an 8-hour average A-weighted sound level, the limit is 90 dB when the noise is relatively constant.

1.4.4 Australia

The maximum permissible A-weighted sound level envisaged (1973) by law is 90 dB at present and 85 dB in 5 years time for continuous exposures (40 hours a week). It also states that no worker may be exposed to A-weighted sound level in excess of 115 dB at any time.

1.4.5 Argentina

The regulations pertaining to noise exposure in Argentina is given in Table 1.4.

1.4.6 France

In 1971, the Technical Committee in the Ministry of Public Health adopted the following limits: a warning limit of 85 dB for the 8-hour average A-weighted sound level and 90 dB as a limit above which a danger is present and there is a real risk of occupational deafness.

1.4.7 Canada

The Labor Code Regulations for noise control in January 1973 states that no employer should allow an employee to work in a place where A-weighted sound level reaches 90 dB or more. The employer should have reasonable grounds

TABLE 1.5

Noise Standards in Canada

A-weighted sound level (dB)	>87 and ≤90	>90 and ≤92	>92 and ≤95	>95 and ≤97	>97 and ≤100	>100 and ≤102	>102 and ≤105	>105 and ≤110	>110 and ≤115	>115
Exposure time (hours/day)	8	6	4	3	2	1½	1	1/2	1/4	0

for not complying with this general provision. In such cases he can authorize his employees to work in their daily jobs as given in Table 1.5. Exposure to A-weighted impulsive noise should not be allowed to exceed 140 dB.

1.4.8 Poland

The decree enacted in 1959 lays down the limit of A-weighted sound level in places, depending on the kind of work done, as

1. 40 dB, for work demanding a high degree of concentration
2. 50 dB, for workshops and other places where high-precision work is required
3. 90 dB elsewhere

Where speech is to be possible, the A-weighted sound level should not exceed 70 dB. Furthermore, the regulations recommend that an attempt be made to keep noise 5 dB below the permissible maximum.

1.4.9 Former Yugoslavia

The regulations (1971) concerning protection against noise in places of work lay down maximum permissible level of 90 dB for the 8-hour average A-weighted sound level.

1.4.10 Former USSR

The Soviet law concerning maximum permissible sound level in occupational settings for 8-hours shift exposure (after EPA 1971) is given in Table 1.6.

1.5 Instrumentation for Noise Measurement

Several instruments are available to measure sound level in the field such as sound level meter, noise dosimeter, octave band analyzer, etc. The selection of a particular instrument basically depends on the intended purpose.

TABLE 1.6

Noise Standards of Former USSR

Situations	A-weighted Sound Level (dB)
1. Ordinary workplaces in factories, etc.	85
2. Laboratories with noise source	75
3. Remote control and observation stations in factory automated process	65
4. Offices with office machinery	60
5. Office where thinking work demands high level of concentration	50

For instance, the ordinary sound-level meter measures only the sound level at a place. Noise dosimeter finds application when the operator's noise exposure is to be measured. Frequency analyzer is used to get a noise spectrum giving sound levels corresponding to various frequency bands of interest. Such noise spectrum is used as the characteristic of the noise source. An important purpose of the noise spectrum is to identify the high-noise-producing components of any large equipment and hence help in the noise abatement process.

1.5.1 Sound Level Meter

The sound level meter is the basic instrument for noise measurement. It consists of a microphone and an electronic circuit, including an attenuator, an amplifier, frequency response network (weighting network), RMS (root mean square) rectifier, and an indicating meter. It measures the RMS sound pressure level in decibels, which is proportional to the intensity or sound energy flow. The weighting networks provided on standard sound level meters are in an attempt to duplicate the response of the human ear to sounds of various loudness levels. These weighting networks cause the sensitivity of the meter to vary the frequency and intensity of sound like the sensitivity of human ear. Action of the indicating meter can be selected as "fast" or "slow." In slow speed, the time constant is 1 second (1000 ms). This is the slowest and is commonly used in environmental noise measurements. In fast speed, the time constant is ⅛ s (125 ms). This is a less commonly used weighting, but it will detect changes in sound level more rapidly. For impulse speed, the time constant is 35 ms for the rise and 1.5 seconds (1500 ms) for the decay. Relatively steady sounds are easily measured using the "fast" response. Unsteady sounds can be averaged with the "slow" response.

1.5.2 Noise Dosimeter

A noise dosimeter is an instrument for assessing the risk of hearing damage. The instrument gives an accurate indication of the total noise exposure the wearer receives throughout the working shift. It automatically takes into account the different sound levels encountered both close to the machinery and at other places where the persons may be while working and during "rests." For measurement over a full working shift, a count of zero percent will be indicated when the wearer will be subjected to a recommended limit of 90 dB(A) for the 8-hour average. A count greater than 100% indicates that the maximum noise dose has been exceeded.

1.5.3 Frequency Analyzer

A frequency analyzer can be used as a simple sound level meter, octave band analyzer, and data logging sound level meter. It is a handheld instrument with simple user interface. For accurate measurements over a long duration, the instrument may be mounted on a tripod.

The instrument can perform both octave as well as one-third octave band analysis, depending on the requirement. It has a large dynamic range and can be set for slow, fast, and impulse RMS detectors. Sufficient memory is available in the instrument to store octave spectra, one-third octave spectra, and simple sound level measurements. The instrument works on a battery and can also be connected to the mains with suitable adapter. A serial interface is also incorporated in the system, which helps the system to get connected to a computer or to a printer for taking direct printouts.

The instrument is equipped with a microphone mounted on a preamplifier. The microphone and preamplifier assembly may or may not be directly mounted on the sound level meter. The incoming sound is first converted to an electrical signal by the microphone. This electrical signal is filtered, detected, and in turn sampled by an analog-to-digital converter. The processor then calculates all the desired data based on the sample data.

1.6 Frequency Analysis of Noise

The amplitude of sound at various frequencies is called *frequency spectrum of noise*. It is given by a set of numbers that describe the amplitude at each frequency or band of frequencies. The procedure for measurement of sound pressure level at different ranges of the frequency spectrum is known as *frequency analysis*. The frequency-cum-sound level distribution helps in locating the frequency bands contributing most to the overall sound level of a source. With this information, it is easier to devise control measures to reduce the high sound level portions of the spectrum.

This instrument is called a "frequency analyzer" or "octave band analyzer." When more specific information about the sound pressure spectral distribution is desired than that provided by octave bands, then narrow band analyzers may be used. A one-third octave band analyzer gives this narrow band analysis and is often required for detailed analysis of machine noise.

A precision sound level meter/frequency analyzer consists of frequency band pass filters. These filters divide the frequency spectrum of sound or vibration into two parts. One part is unchanged, and another part is filtered out. These filters can be either low-pass filters, high-pass filters, or band-pass filters.

The low-pass filter permits signals to pass through that have frequencies below a certain fixed frequency called an *upper cut-off frequency* (F_u). It is therefore used to discriminate against higher frequencies.

The high-pass filter permits signals to pass through that have frequencies above a certain fixed frequency called the *lower cut-off frequency* (F_l). It is therefore used to discriminate against lower frequencies.

A frequency filter that permits signals to pass through that have frequencies above a lower cut-off frequency (F_l) and below an upper cut-off frequency (F_u) is called a *band-pass filter*. The difference between these cut-off frequencies is called the *bandwidth*. The geometric mean frequency is called a *midband frequency* (F_o)

$$F_o = \sqrt{(F_l \cdot F_u)}$$

A bandwidth is represented by its midband frequency.

The upper cut-off frequency (F_u) is related to the lower cut-off frequency (F_l) for frequency analyzers as

$$F_u = K\,F_l$$

where K = constant relating to type of analyzer.

The value of K for different types of analyzers is given in Table 1.7.

The one-third octave band is commonly used for frequency analysis. The lower, upper, and midband frequency for standard one-third octave bands are given in Table 1.8. A special frequency filter that adjusts the amplitude of all parts of the frequency spectrum of the sound is called *frequency filter-weighted*. It can be A, B, C, or flat weighting filter. The A-weighting filter adjusts the amplitudes of a frequency spectrum in the same way the human ear does when exposed to low levels of sound. This weighting is most often used for evaluation of environmental sounds. The B-weighting filter adjusts the levels of a frequency spectrum in the same way the human ear does when exposed to higher levels of sound. This weighting is seldom used.

TABLE 1.7

K-values for Different Analyzer Types

Analyzer Type	K
Octave band	2.000
½ Octave band	$(2)^{1/2} = 1.414$
⅓ Octave band	$(2)^{1/3} = 1.260$
Nth Octave band	$(2)^{1/n}$

TABLE 1.8

Standard One-Third Octave Bands

Lower Frequency (F_l)	Midband Frequency (F_o)	Upper Frequency (F_u)
22.4	25.0	28.2
28.2	31.5	35.5
35.5	40.0	44.7
44.7	50.0	56.3
56.3	63.0	70.9
70.9	80.0	89.2
89.2	100.0	112.0
112.0	125.0	141.0
141.0	160.0	178.0
178.0	200.0	224.0
224.0	250.0	282.0
282.0	315.0	355.0
355.0	400.0	447.0
447.0	500.0	563.0
563.0	630.0	709.0
709.0	800.0	892.0
892.0	1000.0	1120.0
1120.0	1250.0	1410.0
1410.0	1600.0	1780.0
1780.0	2000.0	2240.0
2240.0	2500.0	2820.0
2820.0	3150.0	3550.0
3550.0	4000.0	4470.0
4470.0	5000.0	5630.0
5630.0	6300.0	7090.0
7090.0	8000.0	8920.0
8920.0	10000.0	11200.0
11200.0	12500.0	14100.0
14100.0	16000.0	17800.0
17800.0	20000.0	22400.0

The C-weighting filter adjusts the levels of a frequency spectrum in the same way the human ear does when exposed to high levels of sound. This weighting is most often used for evaluation of equipment noise. The flat-weighting filter is one that does not adjust the levels of a frequency spectrum. Table 1.9 indicates the weighting network frequency response to A, B, and C weighting networks.

TABLE 1.9

Weighting Network Frequency Response to A, B, and C Weighting Networks

Midband Frequencies (Hz)		Weighting Network Frequency Response		
1/3 Octave	**1/1 Octave**	**A**	**B**	**C**
20.0		–50.4	–24.2	–6.2
25.0		–44.7	–20.4	–4.4
31.5	31.5	–39.4	–17.1	–3.0
40.0		–34.6	–14.2	–2.0
50.0		–30.2	–11.6	–1.3
63.0	63.0	–26.2	–9.3	–0.8
80.0		–22.5	–7.4	–0.5
100.0		–19.1	–5.6	–0.3
125.0	125.0	–16.1	–4.2	–0.2
160.0		–13.4	–3.0	–0.1
200.0		–10.9	–2.0	0.0
250.0	250.0	–8.6	–1.3	0.0
315.0		–6.6	–0.8	0.0
400.0		–4.8	–0.5	0.0
500.0	500.0	–3.2	–0.3	0.0
630.0		–1.9	–0.1	0.0
800.0		–0.8	0.0	0.0
1000.0	1000.0	0	0.0	0.0
1250.0		0.6	0.0	0.0
1600.0		1.0	0.0	–0.1
2000.0	2000.0	1.2	–0.1	–0.2
2500.0		1.3	–0.2	–0.3
3150.0		1.2	–0.4	–0.5
4000.0	4000.0	1.0	–0.7	–0.8
5000.0		0.5	–1.2	–1.3
6300.0		–0.1	–1.9	–2.0
8000.0	8000.0	–1.1	–2.9	–3.0
10000.0		–2.5	–4.3	–4.4
12500.0		–4.3	–6.1	–6.2
16000.0	16000.0	–6.6	–8.4	–8.5
20000.0		–9.3	–11.1	–11.2

References

Aljoe, W.W., Bobick, T.G., Redmond, G.W., and Bartholomae, R.C. (1985). "The Bureau of Mines noise control research program—a 10-years review." U.S. Bureau of Mines, I.C. No. 9004, pp. 1–85.

Archibald, R. Mcl. (1964). "Men and machines, the medical problem of the sixties—the effect of mechanization on the pattern of accidents." *The Mining Engineer*, November, pp. 118–124.

Barker, A.H., and Gill, W.F. (1972). Noise, the Symposium on Environmental Engineering in Coal Mining, organized by the Institution of Mining Engineers, 3, Grosvenor Crescent London, 31st October to 2nd November, pp. 191–199.

Bhatia, S.C. (2002). *Handbook of Industrial Pollution and Control*, Vol. 1, Chapter 6. CBS Publishers and Distributors, New Delhi.

Broner, N. (1978). "The effects of low frequency noise on people—a review." *J. Sound Vibrat.*, 58(4), 483–500.

Daniel, W.E., Fulton-Kehoe, D., Smith-Weller, T., and Franklin, G.M. (1998). "Occupational hearing loss in Washington State, 1984–91: II Morbidity and associated costs." *Am. J. Ind. Med.*, 33, 529–536.

Das, J., and Deshbhratar, R.S. (1982). "Noise—a source of environmental pollution in mines." *Indian Min. Eng. J.*, May, pp. 11–15.

Davis, H. (1947). *Hearing and Deafness*. New York, Murray Hill Books, Chapter 3.

DGMS (1975). DGMS Circular No. DGMS (Tech)/18. Directorate General of Mines Safety, Dhanbad, India.

DGMS (1990). DGMS Circular No. DGMS (Tech)/5. Directorate General of Mines Safety, Dhanbad, India.

Dubey, A., and Nath, R. (2000). "A study of noise pollution due to opencast coal mining activity." *Minetech*, 21(1), 37–39.

Franken, P.A.(1974). "Community noise." *Industrial Pollution*, edited by N. Irving Sax. Van Nostrand Reinhold Company, New York.

Gerald R. Coonan (1975). "The coal industry and the ultimate in noise." Third Symposium on Surface Mining and Reclamation, NCA/BCR Coal Conference and Expo II, October 21 to 23, pp. 203–220.

Glorig, A. (1958). *Noise and Your Ear*. New York, Grune and Stratton. Chapter 12 and 14.

Harris, C.M. (1979). *Handbook of Noise Control*. Toronto, McGraw Hill. Chapters 20, 23–33.

Hetu, R., Riverin, L., Lalande, N., Getty, L., and St-Cyr, C. (1988). "Qualitative analysis of the handicap associated with occupational hearing loss." *Br. J. Audiol.*, 22, 251–264.

Hetu, R., Getty, L., and Waridel, S. (1994). "Attitude towards co-workers affected by occupational hearing loss. II: Focus groups interviews." *Br. J. Audiol.*, 28, 313–335.

Hockey, G.R.J. (1972). "The effects of noise on human efficiency and some individual differences." *J. Sound Vibration.*, 20(3), 299–304.

Job, R.F.S. (1988). "Community response to noise—a review of factors influencing the relationship between noise exposure and reaction." *J. Acoust. Soc. Am.*, 83(3), 991–1001.

Khuntia, G.S., and Mishra, P.K. (1994). "Noise-management perspective with special reference to SAIL mines." *Proc. Int. Seminar on Environmental Issues in Minerals and Energy Industry, ENVIROMIN-94*, October 24–26, New Delhi, pp. 358–363.

Krichagin, V.J. (1978). "Health effects of noise exposure." *J. Sound Vibrat.*, 59(1), 65–71.

LaBenz, P., Cohen, A., and Pearson, B. (1967). "A noise and hearing survey of earth moving equipment operators." *Am. Ind. Hyg. Assoc. J.*, 28, 117–128.

Leigh, J.P., and Miller, T.R. (1998). "Job related diseases and occupations within a large workers compensation data set." *American Journal of Industrial Medicine*, 33, 197–211.

Margaret A. Hayley (1969). "Industrial noise deafness and compensation." *Can. Min. J.*, September, 52–54.

Miller, James D. (1974). "Effects of noise on people." *J. Acoust. Soc. Am.*, 56(3), 729–764.

Mitra, P.K. (1987). "Noise—the concealed human killer." *Proc. Natl. Seminar on Environmental Pollution and Control in Mining Coal and Mineral-Based Industries*, IIT Kharagpur, February 13 to 15, pp. 85–86.

Mukhopadhyay, S., and Dey, N.C. (1999). "A comprehensive analysis of dust and noise pollution in mines." *J. Mines Met. Fuels*, March, 80–84.

Neely, Keith K. (1967). "Noise and hearing conservation." *Can. Min. J.*, September, 68–72.

Niebel, B., and Freivalds, A. (1999). *Methods, Standards, and Work Design*, 10th edition. New York, McGraw-Hill.

Pal, A.K., and Saxena, N.C. (1999). "Noise pollution in coal mining complexes." A monograph by ENVIS Centre on Mining Environment, Indian School of Mines, Dhanbad, India.

Pandey, P. (1987a). "Noise management in underground mines." *J. Mines Met. Fuels*, December, 495–502.

Pandey, P. (1987b). "Noise and vibration control in mines." *Proc. Natl. Seminar on Environmental Pollution and Control in Mining Coal and Mineral-Based Industries*, IIT Kharagpur, February 13 to 15, pp. 162–178.

Park, J.E., and Park, K. (1983). *Preventive and Social Medicine*. M/s Banarasidas Bhanot, Jabalpur (India), 173–175.

Peters, T., and Bochum (1978). "Effect of noise on man." Glauckauf + Translation, 114(19–24), 515–516.

Powell, W.H. (1956). "The assessment of noise at collieries." *Trans. Inst. Min. Engineers*, 116, 22–42.

Pujara, K.K. (1998). "Education for noise control." *Proc. 28th Annual Convention of Indian Society of Technical Education*, IIT Delhi, November, pp. 21–23.

ReVelle, C., and ReVelle, P. (1974). "*Source Book on Environment: The Scientific Perspective*. Boston, Houghton Mifflin.

Roychowdhury, S.N., and Dhar, B.B. (1982). "Noise—an occupational hazard and a source of environmental pollution." *Indian Min. Eng. J.*, May, 5–9.

Rudmore, W. (1958). "Hazards of noise exposure." *Noise Control IV*, September, 39–58.

Singh, P.K., Loveson, V.J., and Singh, T.N. (1997). "Impact of noise pollution on workers health of cement plant." Conference on Mines Safety and Mechanization in Surface Mines (MINESAFE) organized by the *Indian Mining and Engineering Journal*, Department of Mining Engineering, REC Rourkela, and Institution of Engineers, India, November 28 to 29, pp. 31–34.

Singh, R., Jaiprakash, K.C., and Singh, T.N. (1999). "Problem of noise pollution in and around industrial area and its abatement." *Indian Min. Eng. J.*, June, 36–40.

Stanley E. Forshaw (1969). "Noise exposure and hearing protection." *Can. Min. J.*, September, 49–52.

Suter, A.H. (2002). "Construction noise: exposure, effects and potential for remediation—a review and analysis." *Am. Ind. Hyg. Assoc. J.*, 63(6), 768–790.

Tiwari, S.N. (1992). "Noise pollution in mines—in brief." *Indian Min. Eng. J.*, November, 21–22.

Tripathy, D.P., and Patnaik, N.K. (1994). "Noise pollution in opencast mines—its impact on human environment." *Proc. Int. Seminar on Impact of Mining on the Environment*, India, 55–65.

Walker, A. (1963). "Noise—its effect and control in mining operations." *Can. Min. Metall. Bull.*, November, 820–834.

Williams, D.A., and Ross, C.R. (1968). "Effects of environmental noise." *Can. Min. J.*, October, 91–93.

Wright, W.C. (1964). *Disease of Workers: Translation from the Latin Text De Morbis Artificum*. New York, Hafner Publishing.

2

Application of Noise Monitoring for Mining Equipments

2.1 Sources of Noise in Heavy Earth-Moving Machinery (HEMM)

A number of studies carried out on noise measurement and its control in the past reflect bulldozers, dumpers, and front-end-loaders as the major sources of noise in open-pit mines (Daniel et al., 1979; Bartholomae and Bobick, 1983; Aljoe et al., 1985 and Valoski, 2000). The above-mentioned studies also indicated major sources of noise in diesel-powered equipments as the exhaust, cooling fan, engine, air inlet, transmission, final drive, and the main frame. Noise from these sources reach the operator via airborne as well as structure-borne paths.

The main source of noise in bulldozer is the diesel engine. The other sources that contribute to the overall noise of dozer equipment are the cooling fan, exhaust, transmission, drive train, and the hydraulic system. Apart from this, for an operating dozer, a significant amount of noise is also produced from the crawler tracks. Typically, the A-weighted sound level of crawler dozers may range from 95 to 105 dB (MSHA, 1983).

The front-end loader ranks second to the bulldozer as a noise offender in the surface mining industry (Bartholomae and Aljoe, 1979). These machines produce A-weighted sound level at operator's position in the range of 95 to 102 dB (MSHA, 1983). The major sources of noise in these machines are the exhaust, engine, and the hydraulic system.

In dumpers, the major source of noise is the diesel engine situated on its right side. The next important source of noise in these machines is the exhaust. The exhaust gas goes out from the rear end. The maximum sound however occurs when the dumper starts its engine, characterized by a sharp, high pitch sound of short duration (Aljoe et al., 1985).

In case of the rope shovel, the motors are the major sources of noise. The motor noise is manifested in two forms, that is, due to aerodynamic and structural vibrations (Harris, 1979). The aerodynamic sources include siren tone, turbulence noise, and whistling tone. The structural vibration is mainly due to mechanical sources (bearing, rotor shaft unbalance, brush and brush holder), and electromagnetic sources (air gap flux, dissymmetry, torque oscillation, magnetostriction). The major motors in a rope shovel are the crowd, the hoist, the propel, and the swing motor. There are generally two hoist motors, two swing motors, and two propel motors all of which are DC motors, whereas the crowd motor is an AC motor. The major noise can be found inside the machinery house where hoist drum reels are present. Here, in addition to motors, various cooling fans also produce high sound levels. The major source of noise in a hydraulic shovel is the engine. The hydraulic shovel has a four-stroke diesel engine. Knocking is an important source of noise in the engine. Knock originates from the phenomenon of auto-ignition of the fuel. In compression ignition engines, uncontrolled self-ignition occurs at the beginning of the pressure rise, giving rise to knock.

Major sources of noise in electrically driven drills include the engine, the compressor, the compressor motor, the hydraulic system, the drill motor, blow air at the point of drilling, the rattling of drill pipes, the fan, and the vibration of the machine. The hydraulic and the electric motors produce noise especially when the drill is in operation. Behind the drill and away from the operator, high sound level is produced from the compressor. The A-weighted sound level to the operators using such machines may range from 75 to 95 dB (MSHA, 1983). In case of diesel-powered rotary drills, the main sources of noise are the diesel exhaust, the fan, and the hydraulic system. These machines produce A-weighted sound levels in the range of 85 to 100 dB at the operator's position (MSHA, 1983).

2.2 Diesel Engine Noise—Characteristics and Control

The most important study on the noise of diesel engine injection equipment was reported by Priede (1967). The origin of noise from two major components of the diesel equipment, namely, the injection pump and the injectors, was discussed in detail. It was shown that the predominant noise of the injection pump originates from the vibration of the pump camshaft. This vibration is due to rapid changes of fluid pressure in the pump elements. This noise was said to lie in the frequency range of 500 to 800 Hz. The injector noise was said to lie around the frequency of either 2 kHz or 5 kHz. It was further said that injection equipment noise is significantly affected by the mounting system. Therefore, the characteristics of engine structure play a significant role in determining the level and the character of the noise emitted by the injection equipment. About 20 dB and 25 dB noise reduction of the injection pump and injector system, respectively, were reported by relevant changes in the design of the components, the engine structure, and the mounting arrangements.

Priede (1971) carried out another study on the origins of automotive vehicle noise. The characteristic of noise produced by major elements of road vehicle was discussed. It was indicated that the noise of engine cooling fan can be tonal as well as broadband in character. This noise is related to the engine speed as:

dB(A) α 55 log N, where N is the engine speed in rev/min.

The diesel engines contribute noise mainly due to combustion and mechanical sources within the engine itself. The noise from these two sources is manifested in two forms, that is, the casing radiated noise and the exhaust noise (Priede, 1967; Priede, 1971; Waters, 1971; Bennett, 1985). It was reported by Priede (1967, 1971) that the casing radiated noise is broadband in character and important over the frequency range 0–5 kHz. It was also said that such noise depends on engine speed and for indirect injection engines (IDI) it follows the law

dB(A) α 40 log N, where N is the engine speed in rev/min.

This statement was further substantiated by Bennett (1985). It was also stated that the noise radiated from the casing is unaffected by the engine load. An increase in A-weighted sound level of the order of 2 dB was reported from no load to full load. It was also reported that the casing-radiated noise does not depend on the engine size. An important conclusion was thus made that larger and slower engines are quieter than the smaller and faster running engines generating identical power.

Priede (1971) also indicated the exhaust noise of road vehicles to depend on engine speed according to the relation

dB(A) α 45 log N, where N is the engine speed in rev/min.

It was said that the exhaust noise is load sensitive and can produce an increase in A-weighted sound level of the order of 15 to 20 dB on an unsilenced exhaust. Based on these investigations, the basic origin of noise with the importance of their contribution to noise inside and outside the automotive vehicles were classified, which is given in Table 2.1.

The effect of operating parameters on sources of vehicle noise was also discussed by Priede (1975). The study indicated significant variation of the order of 9 dB in A-weighted sound levels between engines of different manufacturers. It was further stated that higher H.P engines would be only slightly noisier than smaller and medium H.P engines, if both run at the same speed.

TABLE 2.1

Origin of Noise in Automotive Vehicles

Origin of Noise	Noise Inside the Vehicle	Noise Outside the Vehicle
Engine vibration	Major source of low-frequency noise	Not important
Engine airborne noise and its transmission	Major source of high-frequency noise	Major source of high-frequency noise
Engine exhaust	Not important	Major source of low-frequency noise
Engine inlet	Not important	Major source of low-frequency noise following exhaust
Fan noise	May be noticeable	Can be significant in low- and middle-frequency range
Road-excited vibration	Major source of low-frequency noise	Insignificant
Road-excited tire noise	Insignificant	Significant

Source: Priede, T. (1971). "Origins of automotive vehicle noise." *J. Sound Vibrat.*, 15(1), 61–73.

A study carried out by Jenkins (1975) on analysis and treatment of diesel engine noise indicated a number of noise reduction methods, which could be incorporated at the design stage itself. The study also showed the engine noise to vary with load and speed. Mugridge (1976) suggested various steps to be taken to reduce the noise of cooling fan used in heavy automotive vehicles. It was suggested to reduce the noise of cooling fans by maintaining low axial velocity through the radiator and low fan speed. However, these should be consistent with the pressure head required.

Some studies have been reported to characterize the noise from individual sources for underground machines. As for instance, Bennett (1985) suggested identifying the principal sources of noise along with their characteristic frequency before undertaking any noise control program. It was said that the casing radiated noise in underground diesel locomotives is important over the frequency range 0 to 5 kHz. The noise from cooling fan was said to be tonal as well as broadband in character. The wheel rail noise in case of locomotives/manriding was said to be broadband and important over the frequency range 0 to 5 kHz. Sengupta (1990) described an experimental approach to identify the different noise sources in a load–haul–dump vehicle used underground. A cover-and-expose approach was employed wherein the major sources were covered with lead sheets, and the exhaust was piped away. This gave a baseline blanket condition. In the second stage, the individual components were exposed in order to measure their contributions to the noise level of the machine. Operating conditions of high idle, torque converter stall, and operation in fourth gear were used to accentuate the various sources. In all the cases, noise levels were measured at the operator's position.

2.3 Noise Control of Mining Equipment

In view of the critical nature of the problem, a number of investigators had experimentally characterized noise emissions from mining equipment and evaluated the efficacies of various retrofit treatments aimed at noise control. Noise control of machines can be broadly carried out by two methods:

1. Retrofit approach
2. Maintenance approach

2.3.1 Retrofit Approach

The retrofit approach of noise control for a machine means to put a new part or new equipment in it after it has been in use for some time.

Borland (1972) found that insulating the cabins with fiberglass in trucks can bring down the sound level within the permissible limits. It was also said that the exhaust noise of the engine in trucks can be reduced by taking the exhaust through the body of the trucks. Cummins and Given (1973) and Schwerdtfiger (1978) suggested some important aspects of noise control of machines at design stage itself.

Daniel et al. (1979) evaluated the percentage of surface mine operators overexposed to noise from various types of HEMM and reported the following results: dozers—24%, loaders—8.6%, haulers—3.5%, and diesel-powered shovels and draglines—3.2%. It was found that retrofit noise control treatments applied to dozers could reduce the overall A-weighted sound level at the operator's position by 11.5 dB to 12 dB.

Bartholomae and Aljoe (1979) also carried out a similar retrofit noise control treatment on dozers. It was found that the A-weighted sound pressure level at the operators position of dozers could be brought down to 89–94 dB, which would permit 6 to 8 hours of daily operating time without violating noise regulations. The important retrofit noise control treatments applied on dozers included windshield, canopy absorption, and engine exhaust muffler. The same study also reported retrofit noise control treatment on front-end loaders using exhaust mufflers, seals in the operator's cab and adding sound-absorbing materials on cab surfaces. Following noise control treatment, the A-weighted sound level was reported to reduce by 8 dB with the door of operator's cabin closed. Savich (1982) reported a 17 dB reduction in the A-weighted sound level when the operator's cabin was treated with foam–lead–foam material.

Noise studies carried out by Bartholomae and Bobick (1983) confirm the results of Daniel et al. (1979). The A-weighted sound level was reported to be reduced by 6 dB on dozers by installing a windshield and a muffler and lining the underside of the falling object protective structure (FOPS) with sound absorption material. Sealing all openings and isolating dash and cowling from the vibrating engine reduced the sound level by 5.5 dB. After carrying out all the retrofit treatments, the study revealed an A-weighted sound level of 93 to 94 dB at the operator's position in a dozer with FOPS structure only and 90 dB for dozer with cab. The most important retrofit noise control treatments applied on dozers included windshields, canopy absorption, and engine exhaust mufflers. Some of these retrofit treatments are commercially available, and some of them can be locally fabricated. Similar work was reported by Kovac et al. (1980). A number of retrofit treatments were tried on various mining equipments by United States Bureau of Mines (USBM) and Mine Safety and Health Administrations of USA (MSHA).

Tomlinson (1985) suggested that noise from HEMM could be reduced substantially by using properly designed silencers, and noise absorbent acoustic panels attached to the sides of engine compartments and the radiator of cooling fans. A study carried out by Turner (1986) on scrapers revealed that the A-weighted sound level was reduced by 8 dB upon using exhaust silencers and noise absorbent, damped panels around the engine.

Mukherjee (1987) studied the noise-reducing properties of exhaust mufflers on dozers. The sound pressure levels stayed consistent in the operator's cab, whereas the sound level at different distances from the machine showed some reduction. The authors attributed this finding to the fact that noise from various machine components gets transmitted through the floor and sides of the cab to the operator. Tests were also carried out to study the noise-reducing properties of acoustically treated cabs for front-end-loaders and dozers. Their results indicated 10-dB and 11-dB attenuations of the A-weighted sound level in-cab for front-end-loaders and dozers, respectively.

Some of the retrofit noise control treatments suggested by MSHA (1983) for electric drills include use of muffler at the air inlet of compressor, modification of existing operator cabin or adding a cabin to the existing machine, and modification of the ventilation fans. Maximum noise reduction could be obtained by acoustical modification of the existing cab (MSHA, 1983).

In case of diesel-powered rotary drills, some of the retrofit noise control treatments suggested by MSHA (1983) include use of mufflers at the engine exhaust, addition of acoustic cab, enclosures for engine, modification of existing cab, partial barrier at operator, and modification of the fan with mufflers.

Srinivas and Rao (1989) also reported some positive results of retrofit noise control treatment on dozers and front-end loaders using exhaust silencers, and acoustically treated cabs. Mukhopadhyay and Dey (1998, 1999) suggested using exhaust silencers and fitting of acoustic panels to the rear dump trucks to reduce HEMM noise.

Since pneumatic drills are one of the most important sources of noise in mines, producing A-weighted sound level of the order of 115 to 120 dB at the operator's position, a considerable amount of work had been reported by various researchers pertaining to their noise attenuation (Powell, 1956;

Holdo, 1958; Gordon, 1963; Miller, 1963; Walker, 1963; Chester et al., 1964; Reynolds, 1964; Wallace, 1964; Barker and Gill, 1972; Fischer, 1972; Lemay, 1972; Weber, 1972; Summers and Murphy, 1974; Savich and Wylie, 1975; Visnapuu and Jensen, 1975; Schliesing, 1978; Warner, 1979; Savich, 1982; Das and Deshbhratar, 1982; Roychowdhury and Dhar, 1982; Rainsford and Brown, 1983; Aljoe, 1984; Leahy and Mawson, 1985; Turner, 1986; Vutukuri and Lama, 1986; Aljoe et al., 1987; Stein and Aljoe, 1989; Workman and Davies, 1989; Mukhopadhyay and Dey, 1999). Holdo (1958) reported that of the total delivered useful effect of a compressed air drill under normal operating conditions, the noise energy constitutes only 0.08% of the total energy dissipated. Of this noise energy, 87.5% was said to be due to exhaust with the next largest component being due to impact between piston and drill steel. Similar result was also reported by Walker (1963) and Miller (1963). It was further suggested that efforts to attenuate sound levels of pneumatic drills should concentrate in the frequency range of 0.50 to 0.60 kHz and 1.5 to 7 kHz, as the acoustical energy has most of the sound power in these frequency ranges. The study also revealed that for the frequency range 0.040 to 0.10 kHz, the noise is due to impact between the piston and drill steel and impact between the drill steel and rock; for 0.10 to 2 kHz, the noise is due to exhausting of air from the exhaust ports; and for 2 kHz and above the noise is due to resonance of the steel parts of the drill and resonance of the drill steel itself. A similar result was reported by Gordon (1963) and Savich and Wylie (1975). Later studies by various researchers concentrated on reducing the noise from the exhaust and vibration of the drill rod through retrofit treatments.

Compressors are another important source of noise in mining industry. To reduce the noise exposure of operators and other personnel working nearby a compressor, several studies were carried out (Walker, 1963; Barker and Gill, 1972; Turner, 1986; Pandey, 1987, Das and Deshbhratar, 1982). Most of these studies concentrated on reducing noise using reactive mufflers for attenuating the low-frequency noise at the intake of the compressor. Some studies also made use of a duct so as to remove the intake from the site of the compressor itself. The A-weighted sound level has been reported to reduce by 25 dB at a distance of 1 m from the compressor using reactive silencer (Turner, 1986).

The processing plants are also major source of noise in mines. A considerable amount of work has been reported on reduction of noise in processing plants (Walker, 1963; Thompson, 1972; Watt, 1972; Chedgy, 1976; Chironis, 1978; Rottger, 1978; Leahy and Ridge, 1981; Banerjee, 1982; Roychowdhury and Dhar, 1982; Muldoon and Bobick, 1984; Aljoe et al., 1985; Clements, 1985; Leahy and Mawson, 1985; Turner, 1986; Srinivas and Rao, 1989; Singh et al., 1994; Singh et al., 1997). Most of the work was carried out to reduce the noise from steel plate conveyors, when the material slides on it. These studies also focused on reducing the noise from vibrating screens, ball mill, crusher, chutes, etc.

2.3.2 Maintenance Approach

Maintenance of plant and equipment play a significant role in overall noise control and machinery safety as well as increasing machinery life. Proper maintenance reduces the worn and unbalanced parts of machines that cause both noise and vibration. Parts may become loose, creating more noise because of improper operation or scraping against other parts. Grinding noise may also occur as a result of inadequate lubrication. It is especially important to provide proper maintenance of noise control devices which are added or built into machinery (U.S. Department of Labor, Occupational Safety and Health Administration [OSHA]). If a muffler becomes loose or worn out, it should be fixed or replaced as soon as possible.

Therefore, it is necessary to have a maintenance schedule that covers various aspects like lubricating/oiling all moving parts, balancing, and aligning parts (especially rotating ones), and replacing worn parts. In general, machines get noisier with use due to the following reasons (Worksafe, 1988a):

1. *Worn or chipped gear teeth.* Worn or chipped teeth will not mesh properly. The shiny wear marks are often visible on the teeth.
2. *Worn bearings.* Bearing wear will show up as vibration and noise, squealing from slack drive belts, "piston slap" in motors, air leaks, etc.
3. *Poor lubrication.* This appears as squeaking noises due to friction or excess impact noise in dry and worn gears or bearings.
4. *Imbalance in rotating parts.* Any imbalance in a fan impeller or motor shaft will show up as excess vibration.
5. *Obstruction in airways.* A build-up of dirt or a bent/damaged piece of metal in an airway or near a moving part.
6. *Blunt blades or cutting faces.* Blunt or chipped saw teeth, drill bits, etc., usually make the job noisier as well as slower.
7. *Damaged silencers.* Silencers for air-driven machines or mufflers for engines may become clogged with dirt, rusted, or damaged, so losing their ability to absorb noise.
8. *Removal of a noise-reducing attachment.* Mufflers, silencers, covers, guards, vibration isolators, etc., that reduce noise should never be removed except during maintenance, and then must be replaced.

Bartholomae and Aljoe (1979) indicated that sound level could be substantially reduced by proper maintenance of machines. A study carried out by Central Mining Research Station (1985) reported that periodic maintenance or installation deficiency of HEMM does not contribute significantly to the high sound level emitted by such machines. However, detailed results were not reported in support of this conclusion. Frequency analysis showed that the dominant sound energy was from 100 Hz to 1000 Hz in case of drills, dumpers, and dozers, and in the range of 100 Hz to 2000 Hz in case of scrapers and payloaders. It was emphasized by Berger (1993) that maintenance, periodic adjustment, and replacement of noise control elements are essential for effective noise control.

A study was carried out by Ghosh (1993) to assess the noise emission of HEMM with periodic maintenance schedule. Study was carried out on noise assessment of a few selected HEMM before and after a 1000-hour maintenance schedule. The study indicated considerable noise reduction at various midband frequencies. The equivalent sound pressure level was also reported to decrease on a 1000-hour maintenance schedule. However, a detailed study was not performed to establish the noise reduction at various maintenance schedules.

A study on a total of 11 power tools, including drills, angle grinders, sanders, a circular saw, and a planer, compared the sound levels "before" and "after" routine servicing at a commercial power tool service center in Western Australia by WorkSafe (1988b). The average sound level at 0.5 m distance from all the machines before maintenance was 94 dB(A), and it was reduced to 92 dB(A) after maintenance. The noise reductions were up to 7 dB(A). A similar study on air compressor noise before and after maintenance indicated noise reduction of the order of 7 dB(A) at 1 m distance from the compressor, emphasizing the importance of periodic maintenance. Another study on a reciprocating compressor indicated reduction in A-weighted sound level of the order of 8 dB, by adjusting the valve seat to improve the seal and adding Molyslip additive to the lubricating oil to reduce roughness in the piston stroke. Similarly, the sound level of a pneumatic knife used in abattoirs was reduced by about 8 dB(A) by improving the balance of the rotor vanes and replacing bearings and a worn collar. The sound level of an electric motor and belt drive for an aluminum docking saw was reduced by 15 dB(A), by replacing squealing belts and worn motor bearings and drive pulley bearings. The sound level of ventilation fans in a tunnel was reduced by installing intake and exhaust silencers and by lining the ducts from inside with absorptive fibrous material. This reduced the sound levels from around 85–88 dB(A) down to 81 dB(A) at 1 m distance from the fans.

Maintenance of equipments for the purpose of controlling noise was also emphasized by ECKEL Industries of Canada. It was said that machine maintenance is an important way to control noise.

It was emphasized by Workcover NSW (2004) that a system should be established to ensure regular inspection and maintenance of vibration mountings, impact absorbers, gaskets, seals, silencers, barriers, absorptive materials, and other equipment used to control noise.

It is evident from the previous studies that sound level reductions mostly have been achieved through a variety of retrofit treatments, albeit with varying degrees of success. However, very few studies seem to have addressed the important role played by periodic maintenance of machinery on their noise generation, characteristics, and control. It may be emphasized that for reduction of noise, regular maintenance of the machine is of utmost importance and that the maintenance schedule plays a significant role in noise abatement. In order to achieve maximum noise reduction, maintenance must specifically focus on components generating the highest sound levels. Additionally, periodic maintenance while assisting in noise reduction will also ensure upkeep of machine health.

Investigation was carried out by Vardhan et al. (2006) on the noise characteristics of HEMM (Dumpers and Dozers) used in open-pit mines with various maintenance schedules. Assessment of the sound level after each periodic maintenance revealed major noise reduction after a 1000-hour maintenance schedule. This reduction in sound level was about 6 dB–7 dB for dumpers and 4 dB–5 dB for dozers. At other maintenance schedules, the reduction in noise level was below 1 dB for most of the machines. It was stated that this may be attributed to overhauling of the machine and replacement of certain systems/components on 1000-hours maintenance. A system-wise maintenance study revealed that major noise-generating systems are exhaust, air inlet, cooling, and fuel systems. However, maintenance of lubricant/transmission/hydraulic systems results in no significant improvement in sound level.

2.4 Noise Characteristics of HEMM

2.4.1 Characteristics of HEMM Noise under Loaded and Empty Conditions

HEMM exhibits different noise characteristic under loaded and empty conditions. A study was carried out by Vardhan et al. (2007) on noise assessment of dumpers under idle condition (as under fast-idle condition, the engine noise would be dominant). The measurements were made using a precision sound level meter very close to the dumper body on four dumpers. The difference in the A-weighted equivalent sound pressure level for loaded and empty conditions for all the four dumpers varied from 0.40 to 0.97 dB. This difference in sound level is below 3 dB, and therefore such a difference in the sound pressure level cannot be detected by human ear. The variation of sound levels at nominal one-third octave midband frequencies for all the four dumpers under loaded and empty conditions are shown in Figure 2.1 to Figure 2.4. A clear difference in sound pressure level can be observed at higher midband frequencies. The one-third octave band spectrum indicates that loaded dumpers generate 1 dB to 4 dB less sound pressure level compared to empty ones at higher midband frequencies (greater than 1 kHz). The lower sound pressure level at high frequencies for loaded dumpers is due to the higher mass of the vibrating system (Harris, 1979).

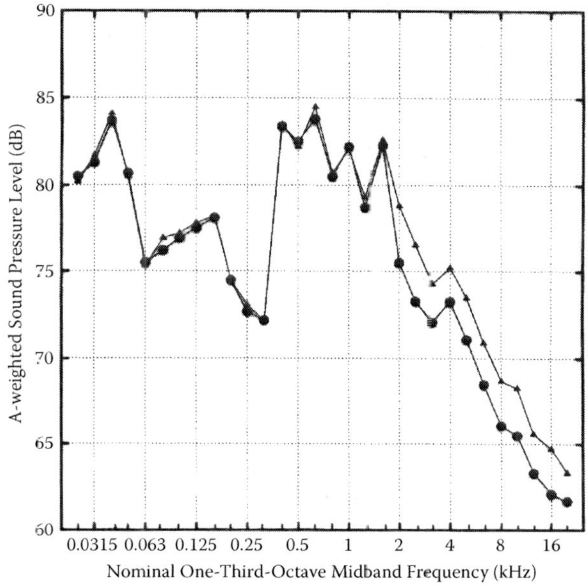

FIGURE 2.1
One-third-octave band spectrum of dumper-1 for empty and loaded conditions: ▲ empty; ● loaded. (After Vardhan, H. et al. [2007]. *J. Inst. Engrs (India)*, 88, August, 41–44.)

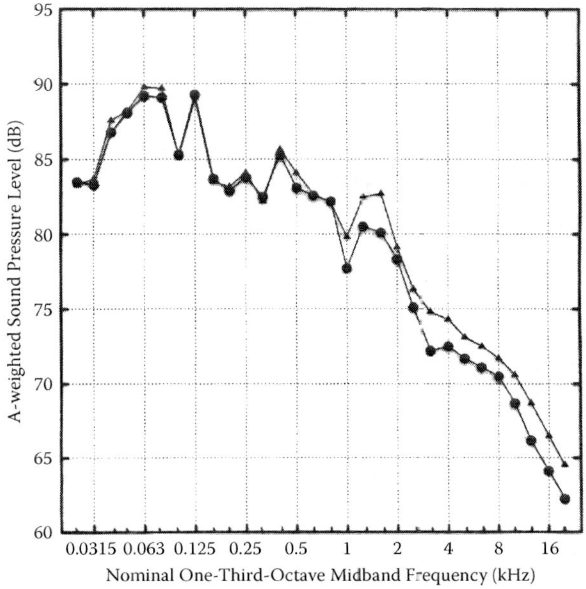

FIGURE 2.2
One-third-octave band spectrum of dumper-2 for empty and loaded conditions: ▲ empty; ● loaded. (After Vardhan, H. et al. [2007]. *J. Inst. Engrs (India)*, 88, August, 41–44.)

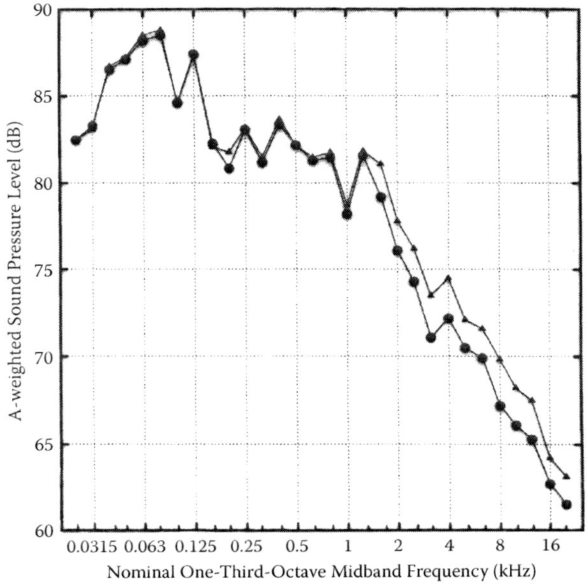

FIGURE 2.3
One-third-octave band spectrum of dumper-3 for empty and loaded conditions: ▲ empty; ● loaded. (After Vardhan, H. et al. [2007]. *J. Inst. Engrs (India)*, 88, August, 41–44.)

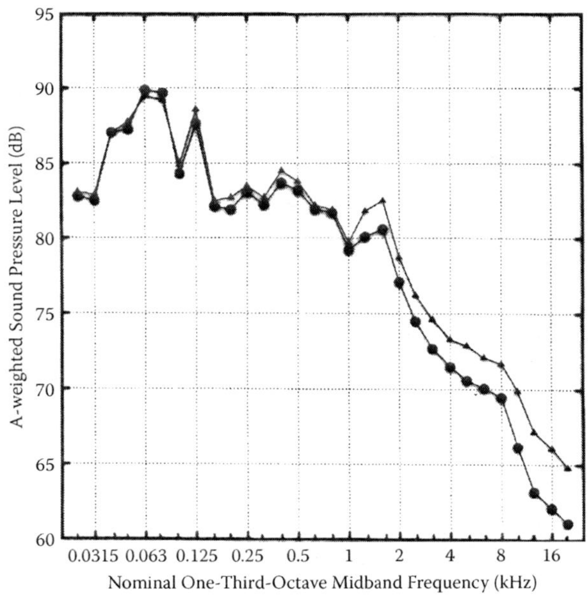

FIGURE 2.4
One-third-octave band spectrum of dumper-4 for empty and loaded conditions: ▲ empty; ● loaded. (After Vardhan, H. et al. [2007]. *J. Inst. Engrs (India)*, 88, August, 41–44.)

2.4.2 Characteristics of Noise Emission from Exhaust, Air Inlet, and Cooling Fan of HEMM

The overall characteristics of HEMM noise is significantly affected by the sound level produced by its individual components. The potential sources of noise in HEMM are the air inlet, exhaust, and cooling fan. Some studies (Bennett, 1985; Sengupta, 1990) have characterized the noise from individual sources for machines used in underground mines and road vehicles. Similar studies for surface mining equipment would enable to characterize the major sources of noise in these machines. The characteristics of noise emanated from HEMM can be studied through frequency analysis. Once the noise characteristics of the components are known, it will be easier to identify the faulty components emitting high sound levels. This will enable the maintenance personnels to take appropriate remedial measures to counter the noise emission from such components.

Vardhan et al. (2005) carried out noise emission studies on dozers and front-end loaders with special emphasis on the noise characteristics of air inlet, exhaust, and cooling fan. The "removal" technique (i.e., isolation of sources) was used to assess the contributions of air inlet, exhaust, and cooling fan noise. Further, an assessment of the performance of an operator's cab was also carried out.

A set of seven test conditions was defined for each earth-moving machine. Table 2.2 lists test conditions D-1 through D-7 for the measurements of the sound from the bulldozers; Table 2.3 lists test conditions L-1 through L-7 for the front-end loaders. For each test condition, the machine was stationary with the transmission in neutral and engine speed maintained at fast idle. Sound pressure levels were measured at the head position of the operator

TABLE 2.2

Test Conditions for Investigation of Two Dozer Sound Sources

(a) Microphone at Operator's Position			
Condition	**Exhaust**	**Cooling Fan**	**Inlet**
D-1	Original muffler	Installed	Original
D-2	New muffler	Installed	Original
D-3	Original muffler	Removed	Original
D-4	Two new mufflers in series + 10 m hose	Installed	Original
D-5	Two new mufflers in series + 10 m hose	Installed	Muffler on air inlet via long flexible hose
(b) Microphone in Engine Compartment			
Condition	**Exhaust**	**Cooling Fan**	**Inlet**
D-6	Original muffler	Installed	Original
D-7	New muffler	Installed	Original

TABLE 2.3

Test Conditions for Investigation of Three Front-End-Loader Sound Sources

(a) Microphone at Operator's Position Inside Cab		
Condition	**Exhaust**	**Inlet**
L-1	Original muffler	Original
L-2	New muffler	Original
(b) Microphone in Engine Compartment		
Condition	**Exhaust**	**Inlet**
L-3	Original muffler	Original
L-4	New muffler	Original
(c) Microphone 3 m in Front of Machine		
Condition	**Exhaust**	**Inlet**
L-5	Original muffler	Original
L-6	Two new mufflers in series + 10 m hose	Original
L-7	Two new mufflers in series + 10 m hose	Muffler on air inlet via long flexible hose

Source: Vardhan, H. et al. (2005). *Noise Control Eng. J.—An International Publication,* American Institute of Physics, 53(2), 37–42.

with the operator present, in the engine compartment, and, for the front-end loaders, at a position 3 m in front of the loader. No other significant sound source was in the vicinity of the machine under test when sound pressure levels were measured. The new muffler mentioned in Table 2.2 and Table 2.3 was a new one of the original design. For test condition D-3, the cooling fan was removed for the brief period required to acquire the data. The period was short enough that no overheating of the engine occurred.

Figure 2.5 shows the changes in the spectrum of the sound at the position of the operator of the bulldozer when a new muffler (condition D-2) was installed in place of the original exhaust muffler (condition D-1). Figure 2.6 shows the effect on the spectrum when the engine was operated with the cooling fan removed (condition D-3). Figure 2.7 shows the effect on the spectrum when the original muffler on the exhaust was supplemented by the addition of two new mufflers in series plus a 10-m-long hose (condition D-4). Figure 2.8 shows the amount of additional noise reduction that was achieved when the sound from the air inlet was suppressed by inclusion of a muffler and a long flexible hose on the inlet duct (condition D-5). Figure 2.9 shows the effect on the spectrum of the sound at the position in the engine compartment when the original muffler was replaced by the new muffler (conditions D-6 and D-7). Figure 2.10 shows the changes in the spectrum

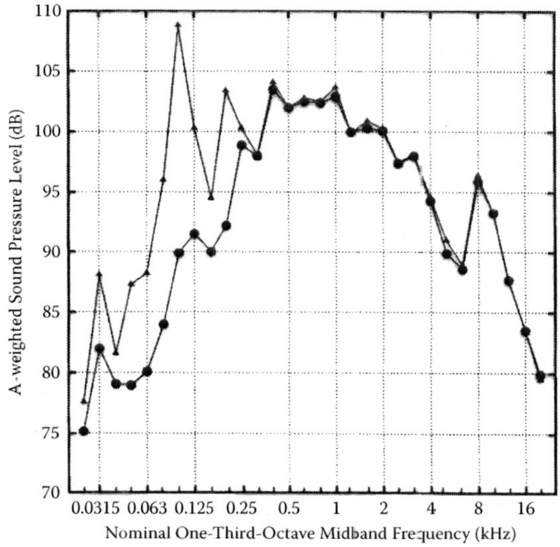

FIGURE 2.5
Effect on A-weighted sound pressure levels at operator's position of replacing the original muffler for the bulldozer-1: ▲ original muffler, condition D-1; ● new muffler, condition D-2. (After Vardhan, H. et al. [2005]. *Noise Control Eng. J.—An International Publication*, American Institute of Physics, 53(2), 37–42.)

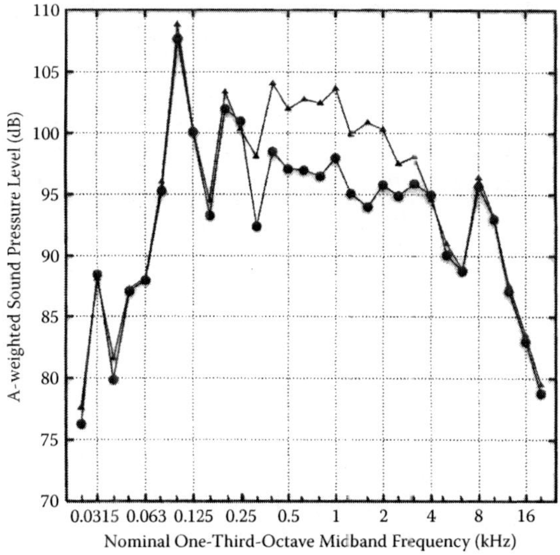

FIGURE 2.6
Effect on A-weighted sound pressure level at operator's position of removing the cooling fan for the bulldozer-1: ▲ fan installed, condition D-1; ● fan removed, condition D-3. (After Vardhan, H. et al. [2005]. *Noise Control Eng. J.—An International Publication*, American Institute of Physics, 53(2), 37–42.)

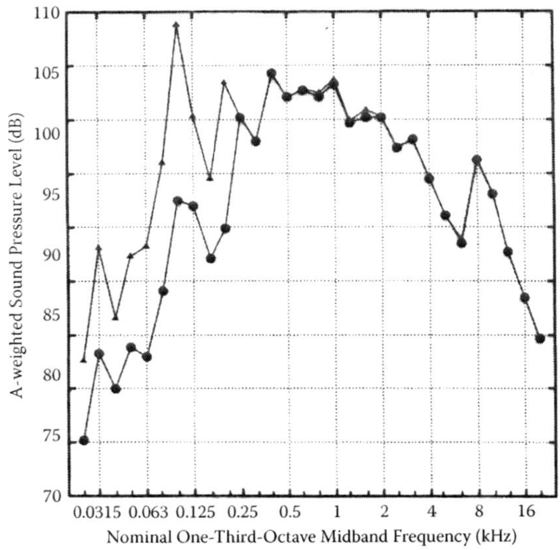

FIGURE 2.7
Effect on A-weighted sound pressure level at operator's position of extensive suppression of exhaust noise for the bulldozer-1: ▲ original muffler, condition D-1; ● exhaust noise extensively suppressed, condition D-4. (After Vardhan, H. et al. [2005]. *Noise Control Eng. J.—An International Publication*, American Institute of Physics, 53(2), 37–42.)

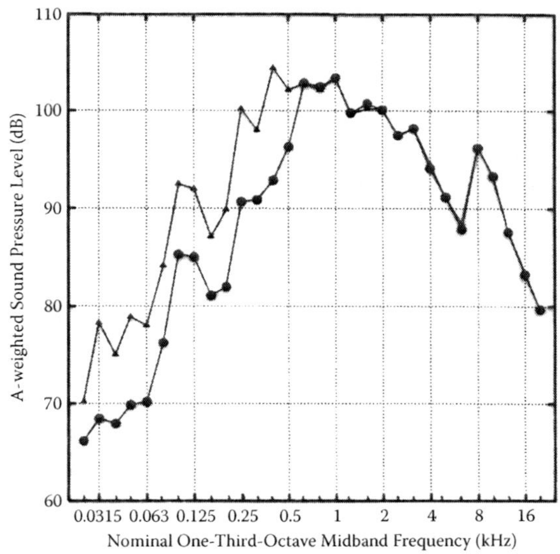

FIGURE 2.8
Effect on A-weighted sound pressure level at operator's position of extensive suppression of exhaust and air-inlet noise from the bulldozer-1: ▲ condition D-4; ● sound from both exhaust and inlet suppressed, condition D-5. (After Vardhan, H. et al. [2005]. *Noise Control Eng. J.—An International Publication*, American Institute of Physics, 53(2), 37–42.)

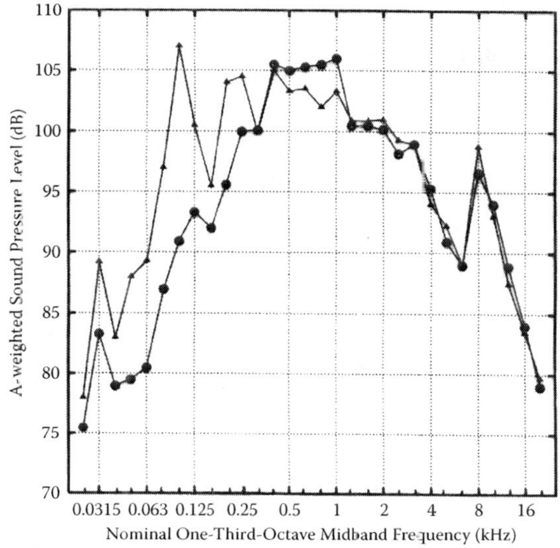

FIGURE 2.9
Effect on A-weighted sound pressure level in engine compartment of replacing the original muffler for the bulldozer-1: ▲ original muffler, condition D-6; ● new muffler, condition D-7. (After Vardhan, H. et al. [2005]. *Noise Control Eng. J.—An International Publication*, American Institute of Physics, 53(2), 37–42.)

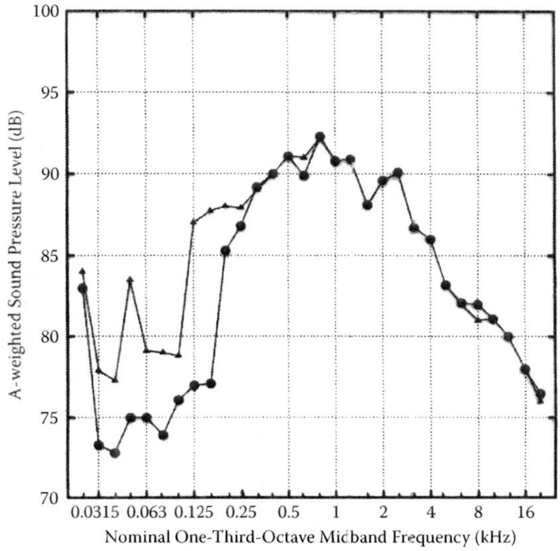

FIGURE 2.10
Effect on A-weighted sound pressure level in operator's cab of replacing the exhaust muffler for the front-end loader-1: ▲ original muffler, condition L-1; ● new muffler, condition L-2. (After Vardhan, H. et al. [2005]. *Noise Control Eng. J.—An International Publication*, American Institute of Physics, 53(2), 37–42.)

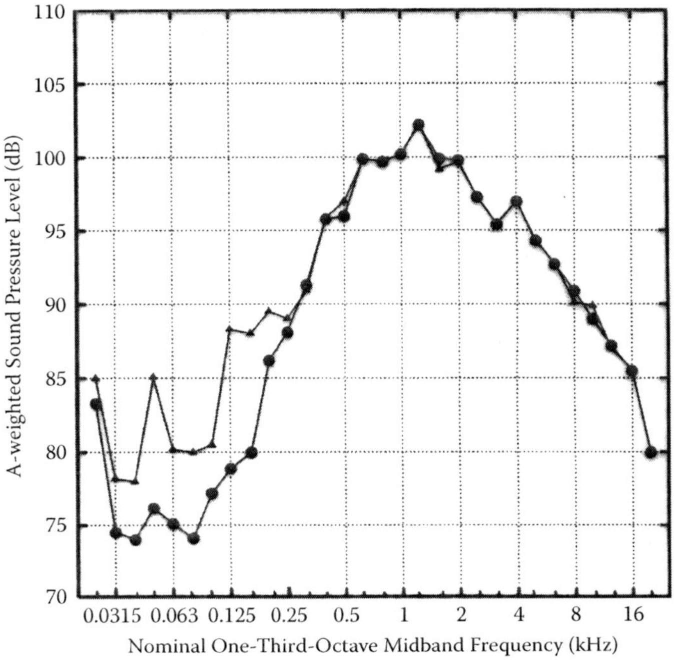

FIGURE 2.11
Effect on A-weighted sound level in engine compartment of replacing the exhaust muffler for the front-end loader-1: ▲ original muffler, condition L-3; ● new muffler, condition L-4. (After Vardhan, H. et al. [2005]. *Noise Control Eng. J.—An International Publication*, American Institute of Physics, 53(2), 37–42.)

of the sound at the operator's position in the cab of the front-end loader when the original muffler was replaced by a new muffler (conditions L-1 and L-2). The door and windows were closed for all measurements of the sound in the cab enclosure.

As was the case for the bulldozers, the noise reductions achieved at the position in the engine compartment of front-end loaders (Figure 2.11) as a result of installing the new exhaust muffler (conditions L-3 and L-4) were similar to those noted at the operator's position (Figure 2.10) and over the same range of midband frequencies. Figure 2.12 shows the spectra in the cab and in the engine compartment (conditions L-1 and L-3) with the original muffler installed for front-end loader. At midband frequencies greater than 400 Hz there was significant noise reduction provided by the enclosure. Figure 2.13 shows the changes to the spectrum of A-weighted, one-third-octave-band sound pressure levels at the position 3 m in front of the loader when the original muffler (condition L-5) was augmented by the addition of two new mufflers in series and a 10-m-long hose (condition L-6) and then with a muffler and hose attached to the air inlet as well as the two new exhaust mufflers and hose (condition L-7). The noise reductions achieved

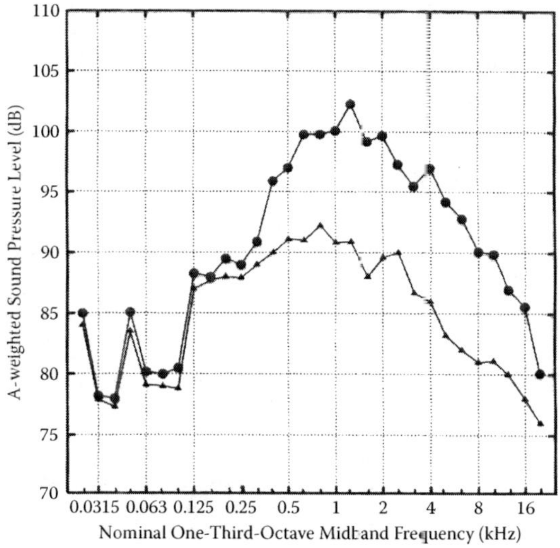

FIGURE 2.12
Effect on A-weighted sound pressure level of the enclosed cab for the operator of the front-end loader-1: ▲ in the cab, condition L-1; ● in the engine compartment, condition L-3. (After Vardhan, H. et al. [2005]. *Noise Control Eng. J.—An International Publication*, American Institute of Physics, 53(2), 37–42.)

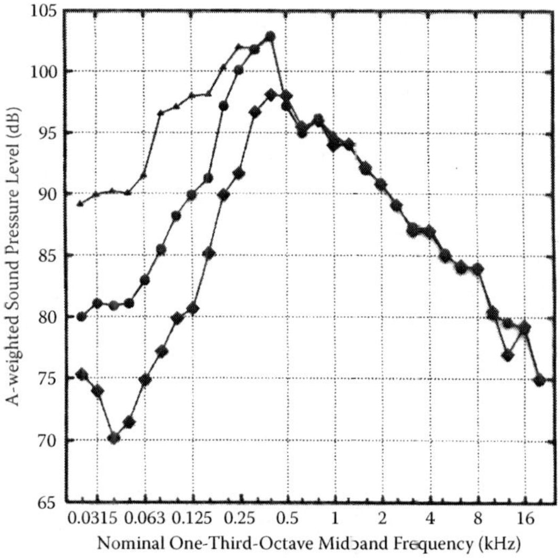

FIGURE 2.13
Effect on A-weighted sound pressure level at a distance of 3 m in front of the front-end loader-1 of extensive suppression of exhaust noise and of exhaust noise and noise from the air inlet: ▲ original muffler, condition L-5; ● exhaust extensively suppressed, condition L-6; ▪ both exhaust and air inlet extensively suppressed, condition L-7. (After Vardhan, H. et al. [2005]. *Noise Control Eng. J.—An International Publication*, American Institute of Physics, 53(2), 37–42.)

by use of these extensive noise reduction techniques were similar to those observed at the operator's position for the bulldozers (Figure 2.8) and indicated that significant reduction in low-frequency sound could be achieved by development of improved noise-control elements for the engine air inlet as well as the exhaust.

It is evident from the above that, the exhaust noise for HEMM is predominant at lower midband frequencies up to 250 Hz, whereas the air inlet noise is predominant at midband frequencies up to 500 Hz for HEMM. The cooling fan noise on the engine of HEMM is significant with midband frequencies from 315 Hz to 3.15 kHz. It is also evident that the operator's cab is likely to be a good attenuator of high-frequency (>400 Hz) noise in HEMM. Therefore, the best method to reduce the operator's exposure to high-frequency noise is by proper design of the cab, making it leakproof. In fact, suppliers of machines generally take care of this. But with passage of time, the cab condition deteriorates and loses its effectiveness in reducing the high frequency noise. Therefore, whenever the machine is sent to workshops for its periodic maintenance, proper attention should be paid to the condition of the operator's cab. The effectiveness of the cab can be assessed by comparing the noise spectrum at the operator's position with that of the noise spectrum of the same machine when newly deployed.

Hence, a well-maintained cab for an operator of heavy earth-moving machines should preferably be air-conditioned or at least well ventilated in order to reduce the need to open the windows while the machines are being operated. In conjunction with improved noise-control elements for the sound from the engine exhaust and air inlet, the operator's noise exposure would be significantly reduced. The use of improved noise-control elements for the engine exhaust and air inlet would also reduce the level of noise experienced by those working in the vicinity of these machines.

References

Aljoe, W.W., Bobick, T.G., Redmond, G.W., and Bartholomae, R.C. (1985). "The Bureau of Mines noise control research programme—a 10-years review." U.S. Bureau of Mines, I.C. No. 9004, pp. 1–85.

Aljoe, W.W. (1984). "Quieted percussion drills." Proceedings: Bureau of Mines Technology Transfer Seminars, Pittsburgh, PA, July 24 and Denver, CO, July 26, U.S. Bureau of Mines, I.C. No. 8986, pp. 75–89.

Aljoe, W.W., Stein, R.R., and Bartholomae, R.C. (1987). "Test apparatus for measuring sound power level of drills." U.S. Bureau of Mines, I.C. No. 9166, pp. 1–35.

Banerjee, S.P. (1982). "Noise, air-blast and vibration problems due to mining." *Proc. of National Seminar on "Minerals and Ecology,"* March, 6–7, Dhanbad, pp. 1–11.

Barker, A.H., and Gill, W.F. (1972). "Noise." "Symposium on Environmental Engineering in coal mining." organised by the Institution of Mining Engineers, 3, Grosvenor Crescent London, 31st October to 2nd November, pp. 191–199.

Bartholomae, R.C., and Aljoe, W.W. (1979). "The noise exposure of mobile machine operators in U.S. surface coal mines and noise control techniques." U.S. Bureau of Mines, I.C. No. 8837, pp. 19–35.

Bartholomae, R.C., and Bobick, T.G. (1983). "Bulldozer noise control." U.S. Bureau of Mines, I.C. No. 8947, pp. 81–86.

Bennett, S.C. (1985). "Noise characteristics of underground transport systems." *Min. Technol.*, March, 95–99.

Berger, E.H. (1993). "The naked truth about NRRs." Available at www.aearo.com.

Borland, R.M. (1972). "Buyers asked for and got noise suppression in trucks." *Can. Min. J.*, 93, September, 67–68.

Clements, J. (1985). "Coal preparation plant noise problems." *Min. Technol.*, 67(773), 85–90.

Central Mining Research Station (1985). "Noise problem arising out of coal mining activities and its abatement." CMRS report, CMRS, Dhanbad, India.

Chedgy, D.G. (1976). "Reduction of environmental noise levels at Meadow River No.1 preparation plant." Society of Mining Engineers, AIME, Transaction, Vol. 260, December, pp. 351–355.

Chester, J.W., Dewoody, R.T., and Miller, W.C. (1964). "Noise from pneumatic rock drills: Shape and exit noise of an exhaust muffler." U.S. Bureau of Mines, R.I. No. 6450, pp. 1–12.

Chironis, N.P. (1978). "Reducing noise in preparation plant." *Coal Age*, February, 164–171.

Cummins, A.B., and Given, I.A. (1973). *SME Mining Engineering Handbook*. Society of Mining Engineers, AIME, New York, pp. 27/84–27/85.

Daniel, J.H., Burks, J.A., Bartholomae, R.C., Madden, R., and Ungar, E.E. (1979). "The noise exposure of operators of mobile machines in U.S. surface coal mines." U.S. Bureau of Mines, I.C. No. 8837, pp. 1–24.

Das, J., and Deshbhratar, R.S. (1982). "Noise—a source of environmental pollution in mines." *Indian Min. Eng. J.*, May, 11–15.

Fischer, H.C. (1972). "Noise control of pneumatic mining equipment." *Can. Min. J.*, 83, October, 75–85.

Ghosh, N. (1993). "Assessment of sound pressure level with maintenance of various HEMM at Rajmahal opencast project." M. Tech. Thesis submitted to Indian School of Mines, Dhanbad, India.

Gordon, J.E. (1963). "Noise suppression on rock drills." *Can. Min. Metall. Bull.*, Vol. 56, November, pp. 835–838.

Harris, C.M. (1979). *Handbook of Noise Control*. Toronto: McGraw Hill Book Co. Inc., Chapters 20, 23 to 33.

Holdo, J. (1958). "Energy consumed by rock drill noise." *Mining Magazine*, August.

Jenkins, S.H. (1975). "Analysis and treatment of diesel engine noise." *J. Sound Vibrat.*, 43(2), 293–304.

Kovac, J.G., Bartholomae, R.C., Bockosh, G.R. (1980). "Bulldozer noise control." *Proc., INTER-NOISE* 80, December, Vol. 1, pp. 457–460.

Leahy, J.C., and Mawson, B. (1985). "Control of noise in UK coal mines." *21st Int. Conf. on Safety in Mines Research Institutes*, Sydney, October, pp. 413–420.

Leahy, J.C., and Ridge, R. (1981). "The application of resilient materials and the reduction of noise in coal preparation plant." *Proc. of the First Australian Coal Preparation Conference*, New Castle.

Lemay, G. (1972). "Campaign quiets worst noise offenders." *Can. Min. J.*, 93, September, 59–60.

Miller, W.C. (1963). "Noise from pneumatic rock drills—measurement and significance." U.S. Bureau of Mines, R.I. No. 6165, pp. 1–30.

Mugridge, B.D. (1976). "The noise of cooling fans used in heavy automotive vehicles." *J. Sound Vibrat.*, 44(3), 349–367.

Mukherjee, R.N. (1987). "Noise an occupational health hazard in industrial activities in India." *Proc. of the National seminar on "Environmental Pollution and Control in Mining Coal and Mineral Based Industries*." IIT Kharagpur, February 13 to 15, pp. 204–212.

Mukhopadhyay, S., and Dey, N.C. (1998). "Noise pollution in mining—a critical appraisal." *Indian Min. Eng. J.*, October, pp. 23–26.

Mukhopadhyay, S., and Dey, N.C. (1999). "A comprehensive analysis of dust and noise pollution in mines." *J. Mines Met. Fuels*, March, 80–84.

Muldoon, T.L., and Bobick, T.G. (1984). "Retrofit noise control for crushing and screening plants." *Proc.: Bureau of Mines Technology Transfer Seminars*, Pittsburgh, PA, July 24 and Denver, Co, July 26, U.S. Bureau of Mines, I.C. No. 8986, pp. 107–119.

Noise control technologies by ECKEL Industries of Canada available at http://www.eckel.ca/art_industrial_noise.html

Pandey, P. (1987). "Noise management in underground mines." *J. Mines Met. Fuels*, December, 495–502.

Powell, W.H. (1956). "The assessment of noise at collieries." *Trans. Inst. Min. Engrs.*, 116, 22–42.

Priede, T. (1967a). "Noise of diesel engine injection equipment." *J. Sound Vibrat.*, 6(3), November, 443–459.

Priede, T. (1967b). "Noise and vibration problem in commercial vehicles." *J. Sound Vibrat.*, 5(1), 129–154.

Priede, T. (1971). "Origins of automotive vehicle noise." *J. Sound Vibrat.*, 15(1), 61–73.

Priede, T. (1975). "The effect of operating parameters on sources of vehicle noise." *J. Sound Vibrat.*, 43(2), 239–252.

Rainsford, C.J., and Brown, J.R. (1983). "The propagation and attenuation of noise in coal mine roadways." The Mining Engineers, Paper No. 4933, September, pp. 115–121.

Reynolds, J.W. (1964). "Noise control at the Sullivan Mine." *Can. Min. Metall. Bull.*, LXVII, 13–16.

Roychowdhury, S.N., and Dhar, B.B. (1982). "Noise—an occupational hazard and a source of environmental pollution." *Indian Min. Eng. J.*, May, 5–9.

Rottger (1978). "Noise reduction for winning and transport operations." *Gluckauf*, 114, December, 525–526.

Savich, M.U. (1982). "Abatement of noise and vibration in the Canadian mining industry." *Can. Min. J.*, 103, August, 31–38.

Savich, M.U., and Wylie, J. (1975). "Noise attenuation in rock drills." *Can. Min. J.*, 96, October, 39–42, 44, 66.

Schliesing (1978). "Noise reduction in roadway drivage." *Gluckauf*, 114, December, 524–525.

Schwerdtfiger (1978). "Main possibilities of noise control with examples of methods used above ground." *Gluckauf*, 114, December, 522–523.

Sengupta, M. (1990). *Mine Environmental Engineering*. Vol. II, Boca Raton, FL, CRC Press, pp. 31–40.

Singh, P.K., Pal, A.K., and Mitra, H. (1994). "Noise pollution in cement plant and control measures—a case discussion." *Proc. Int. Conf. on "Environmental issues in Minerals and Energy Industry" ENVIRONMIN–94*, 24 to 26 October, New Delhi, India, pp. 176–181.

Singh, P.K., Loveson, V.J., and Singh, T.N. (1997). "Impact of noise pollution on workers health of cement plant." Conference on "Mines safety and mechanization in surface mines, MINESAFE." Organized by The Indian Mining and Engineering Journal, Department of Mining Engineering, REC Rourkela and The Institution of Engineers, India, November 28 to 29, pp. 31–34.

Srinivas, P., and Rao, Y.V. (1989). "Assessment and control of noise pollution in mining industry." Fifth National Convention of Environmental Engineers, Madras, India, 23 and 24, September.

Stein, R.R., and Aljoe, W.W. (1989). *Noise Test Report: Hand Held Pneumatic Rotary Drill*. U.S. Bureau of Mines, R.I. No. 9269, pp. 1–11.

Summers, C.R., and Murphy, J.N. (1974). "Noise abatement of pneumatic rock drill." U.S. Bureau of Mines, R.I. No. 7998.

Thompson, P.R. (1972). "Unique noise problem given practical solution." *Can. Min. J.*, 93, September, 58–59.

Tomlinson, J.D. (1985). "Opencast mining and the environment." *Colliery Guardian*, 233, June, 260–263.

Turner, S.K. (1986). "Noise control in South Midlands Area." The Mining Engineer, Paper No. 5030, January, 289–294.

U.S. Department of Labor, OSHA, (2005). "Noise control: A guide for workers and employers." Available at www.nonoise.org/hearing/noisecon/noisecon.htm.

U.S. Mine Safety and Health Administration (1983). "Noise control abstracts." Compiled by MSHA Health and Safety Technology Centers, Denver, CO, and Pittsburgh, PA. Available at www.msha.gov.

Valoski, M.P. (2000). "Noise control in metal/nonmetal mining." *Stone Review*, May/June, 31–34.

Vardhan, H., Karmakar, N.C., and Rao, Y.V. (2007). "Noise assessment of dumpers under loaded and empty conditions." *J. Inst. Engrs (India)*, 88, August, 41–44.

Vardhan, H., Karmakar, N.C., and Rao, Y.V. (2005). "Experimental study of sources of noise from heavy earth moving machinery." *Noise Control Eng. J.*, American Institute of Physics, 53(2), 37–42.

Vardhan, H., Karmakar, N.C., and Rao, Y.V. (2006). "Assessment of heavy earth-moving machinery noise vis-à-vis routine maintenance." *Noise Control Eng. J.*, American Institute of Physics, 54(2), 64–78.

Visnapuu, A., and Jensen, J.W. (1975). "Noise reduction of a pneumatic rock drill." U.S. Bureau of Mines, R.I. No. 8082, pp. 1–23.

Vutukuri, V.S., and Lama R.D. (1986). *Environmental Engineering in Mines*. NewYork, Cambridge University Press, pp. 417–423.

Walker, A. (1963). "Noise—its effect and control in mining operations." *Can. Min. Metall. Bull.*, November, 820–834.

Wallace, A.W. (1964). "Rock drill noise—an approach." *Can. Min. Metall. Bull.*, 57, October, 1038–1041.

Warner, E.M. (1979). "Noise problems with underground mining machinery." *Soc. Min. Eng.*, January, 62–69.

Waters, P.E. (1971). "Commercial road vehicle noise." *J. Sound Vibrat.*, 16(4), September, 627–630.

Watt, A. (1972). "Noise suppression at grinding mills." *Can. Min. J.*, 93, September, 68–69.

Weber, B.H. (1972). "Noise suppression on rock drills." *Can. Min. J.*, 93, September, 63–64.

Workcover NSW (2004). "Noise—management and protection of hearing at work" code of practice by Workcover NSW Health and Safety Code of Practice. Available at www.workcover.nsw.gov.au.

Workman, C.L., and Davies (1989). "Noise and hearing in a trackless mining environment." *J. S. Afr. Inst. Min. Metall.*, July, pp. 211–217.

Worksafe (1998a). Department of Consumer and Employment Protection, Government of Western Australia, "Maintaining your machines." Available at http://www.safetyline.wa.gov.au.

Worksafe (1998b). Department of Consumer and Employment Protection, Government of Western Australia, "Noise control case study: Reduction of noise from a reciprocating compressor through maintenance." available at http://www.safetyline.wa.gov.au.

3

Rock Mechanics: Application of Sound Level

3.1 Introduction

Rocks have been used as construction materials since the dawn of civilization. Different structures have been built on rock, in rock, and have been made of rock, including houses, bridges, dams, tunnels, and caverns. Rocks differ from most other engineering materials in that they contain discontinuities such as joints, bedding planes, folds, sheared zones, and faults that render their structure discontinuous. Hence, determination of physico-mechanical properties of rocks is an important part of rock mechanics.

3.1.1 Determination of Rock Properties

There are different methods available for determining rock properties, which can be divided into two general categories: direct and indirect. The direct methods include laboratory and in situ tests. Many rock mechanics and rock engineering textbooks provide information on conducting laboratory and in situ tests to determine rock properties. In addition, the American Society for Testing and Materials (ASTM) and International Society for Rock Mechanics (ISRM) provide guidance related to the specific procedures involved in performing the actual laboratory and in situ tests.

The direct methods have certain limitations. To obtain realistic results of rock properties, careful preparation of rock samples is required. Because of any number of circumstances, it may prove to be difficult to prepare standard cores from weak, highly fractured, thinly bedded, foliated, and/or block-in-matrix rocks. Weak to very weak rocks may further deteriorate during coring and fail to yield good quality samples. Carrying out such processes would be very difficult, time-consuming, and expensive. Because of the above-mentioned limitations of direct methods, current practice relies heavily on indirect methods such as simple prediction models, which are attractive for engineering geologists (Zhang, 2005).

The indirect methods include empirical or theoretical correlations, combination of intact rock and discontinuity properties using analytical or numerical methods, and back-analysis using field results of prototype observations. The indirect methods can be used not only for determining rock properties but also for checking test results. Data resulting from laboratory and in situ tests are often not completely consistent with other data obtained for a particular project. The indirect methods such as the empirical or theoretical correlations can be used to check the data from tests and investigate the reasons for the inconsistency.

3.1.2 Usefulness of Sound Level in Determining Rock Properties

The drillability of a rock depends on many factors like bit type and diameter, rotational speed and penetration rate of the drill bit. The rock properties such as compressive strength, porosity, density, etc., are uncontrollable parameters during any drilling process. The process of drilling, in general, always produces sound as a by-product. Sound is widely used as a diagnostic tool for identification of faulty components in the mechanical industry. A couple of studies in oil and gas industries have proposed a technique called 'Seismic While Drilling' for estimating rock formations (Asanuma and Niitsuma, 1996; Hsu, 1997). A few studies have proposed the use of noise produced by the bit during drilling as a seismic source for surveying the area around a well (Rector and Hardage, 1992; Miranda, 1996; Tsuru and Kozawa, 1998; Aleotti et al., 1999). The application of sound level produced during drilling, in the mining industry for estimating rock properties has not been much explored. It needs to be emphasized that the Seismic While Drilling technique is different from the technique of estimating rock properties using sound levels produced during drilling.

3.1.3 Artificial Neural Network Modeling

In the engineering domain, algebraic and differential equations are used to describe the behavior and functional properties of real systems and to create mathematical models to represent them. Such approaches require accurate knowledge of system dynamics and the use of estimation techniques and numerical calculations to emulate the system operation. The complexity of the problem itself may introduce uncertainties that make the modeling nonrealistic or inaccurate. Since sound produced during drilling and rock drilling process itself is complicated, Artificial Neural Networks (ANN) can be effectively used to develop a prediction model that may predict the rock properties with precision and accuracy. ANNs have many advantages such as fast prediction responses, noise suppression capabilities, ability to handle large amount of data, and the ability to model complex relationships between the inputs and the outputs without the need for having knowledge about the underlying distributions in the data (Haykin, 1999).

In this book, based on the experimental results, ANN models have been presented for the prediction of rock properties. ANN models presented here were developed using Multi Layer Perceptron (MLP) and Radial

Basis Function (RBF) networks. MLP is a widely used feed-forward neural network type, whereas RBF is a relatively new class of feed-forward network that works on a different principle when compared to MLP (Anderson, 1995).

The ANN modeling presented in this book for predicting rock properties is an effort to comprehensively understand the modeling capability of ANN for these kinds of problems, which will really help in better and faster predictions with the available experimental data.

3.2 Experimental Work on Rock Properties

Assessing the physico-mechanical properties of rock is one of the important factors of concern to the engineers in the general field of rock excavation, especially for performance prediction purposes. Since 1974, and through its commission on testing methods, the International Society for Rock Mechanics (ISRM) has generated a succession of suggested methods for measuring rock properties both in the laboratory and in the field. Some of the laboratory methods are determination of water content, porosity, density and related properties, hardness and abrasiveness of rocks, sound velocity, point load strength, uniaxial compressive strength and deformability of rock materials, shear strength, tensile strength of rock materials, complete stress strain curve for intact rock in uniaxial compression, etc. These tests, along with site characterization and field tests, were compiled and edited by Ulusay and Hudson (2007).

Schmidt designed a portable hammer to conduct nondestructive tests on concrete (Schmidt, 1951). The Schmidt hammer is one of the widely used portable instrument for estimating the rock strength indirectly. It measures the surface rebound hardness of the tested material. Aydin (2009) proposed a revised suggested method, which supersedes the portion of an earlier ISRM document, to determine the rebound hardness of rock surfaces both in laboratory conditions and in situ with an emphasis on the use of this hardness value as an index of the uniaxial compressive strength and Young's modulus of rock materials.

3.2.1 Determination of UCS Using Schmidt Hammer and Equotip L Values

The Schmidt hammer, manufactured by Proceq in Zurich, Switzerland (Proceq, 1977a), measures the distance of rebound of a plunger that is released by a spring toward the surface of a rock. The Schmidt Hammer consists of a spring-loaded piston that is released when the plunger is pressed against a surface (Figure 3.1). The impact of the piston onto the plunger transfers the energy to the material. The extent to which this energy is recovered depends

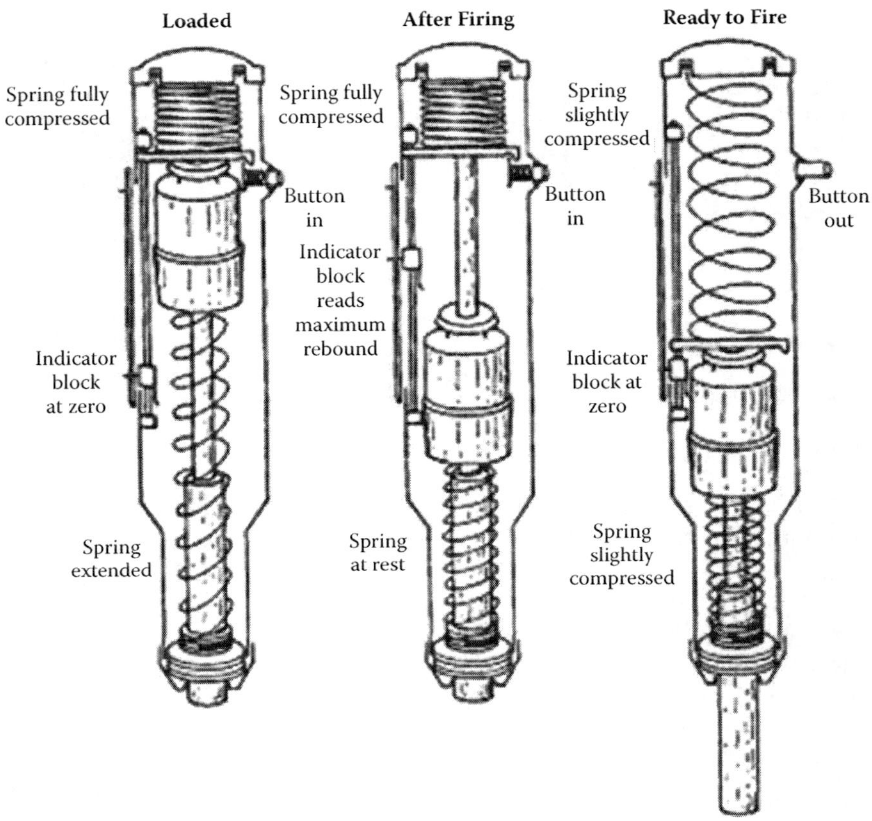

FIGURE 3.1
Working principle of Schmidt Hammer. (From McCarroll, D. [1994]. *Beck Ceditor. Dating in Exposed and Surface Contexts*. Albuquerque: Univ. New Mexico Press, pp. 29–45.)

on the hardness (or impact penetration/damage resistance) of the material, which is expressed as a percentage of the maximum stretched length of the key spring before the release of the piston to its length after the rebound.

The company has also developed a lighter, smaller piece of equipment, the Equotip hardness tester (Proceq, 1977b), which measures the hardness of metallic materials (Figure 3.2). The tester is a relatively new product in the field of hardness testing and was developed to measure the hardness of nonmetallic materials. It is a small battery-operated electronic spring-loaded device, in which a 3-mm-diameter spherical tungsten carbide test tip is mounted in an impact body, which bounces under spring force against a test surface from which it rebounds. A digital display shows the measured values of the impact and rebound velocities. The basic Type D impact device delivers an impact energy of 11 Nmm. Impact devices with an impact energy of 3 Nmm (Type C) and 90 Nmm (Type G) are available. Using samples of various rock

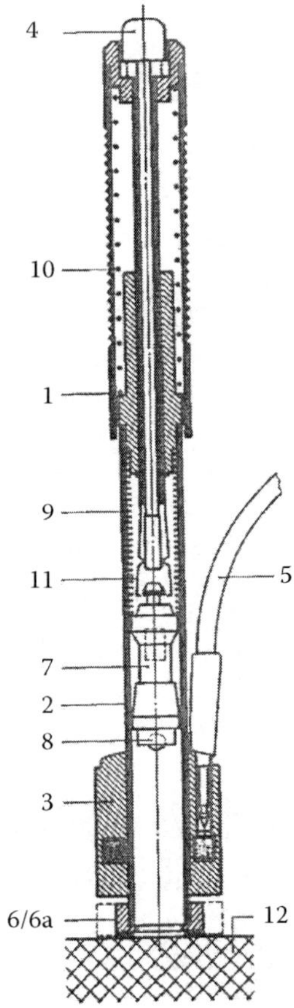

FIGURE 3.2
Schematic design of Equotip hardness tester: (1) loading tube, (2) guide tube, (3) coil with coil holder, (4) release button, (5) connection cable leading to indicating device with coil plug, (6) large support ring, (6a) small support ring (7) impact body, (8) spherical test tip, (9) impact spring, (10) loading spring, (11) catch chuck, and (12) material to be tested.

types, Verwaal and Mulder (1993), investigated the possibility of predicting UCS from the Equotip L-values. A diagram was presented showing the relationship between UCS and L-values. Further, the influence of the surface roughness on the Equotip measurement was also discussed. Kawasaki et al. (2000) considered the use of Equotip testing to establish the strength of rocks in the field. This study was focused on unweathered rocks, and it established the effects of the test conditions, including the size, shape, roughness, and the impact direction. Equotip hardness tests, unconfined compression tests and elastic wave measurements were undertaken by Kawasaki et al. (2002) using cored samples of a number of rock types, including sandstone, shale, greenschist, hornfels, and granite, collected from several locations in Japan.

It was suggested that UCS could be estimated from the Equotip L-values using the relation $UCS = aL + b$, where L is Equotip hardness, a and b are coefficients, depending on rock types.

Szlavin (1974) analyzed the possibility of existence of statistically significant correlations between the mechanical properties of a rock which would enable estimates to be made of one property from any other single property. Various tests such as Compressive Strength, Tensile Strength, Shore Hardness, Indentation, Specific Energy, and Abrasivity were conducted on a number of samples and the arithmetic mean value was calculated and used in the analysis. A program was devised so that the test results could be fed into a computer. The relationships between the variables were obtained in terms of regression coefficients, standard deviations and correlation coefficients. A comparison of the results showed that the majority of the 'direct' mechanical properties, i.e., strength and hardness, can be estimated with reasonable accuracy from each other but greater errors are involved in the determination of, and calculation from, the energy based units, i.e., specific energy index and abrasivity. It was also said that the ratio of uniaxial compressive strength and specific energy is approximately constant. Further, it was also concluded that, National Coal Board's (NCB, England) cone indenter is a suitable instrument for making rapid assessments of rock strength and specific energy.

3.2.2 Relationship between UCS and Indentation Hardness Index

Szwedzicki (1998) proposed a standard indentation test, as a measure of hardness and its use as a predictor of the UCS. The proposed procedure includes application of a standard indenter, specification of a standardized loading rate, criteria for termination of the test, specification for the properties of the cementing agent, and application of continuous data logging. In the test, the indenter was forced against the rock sample, which was cemented in a mould. The standardized indenter is a conical tip having a 60° cone and 5-mm-radius spherical tip. The tip transmits the load to the specimen. The value of the indentation hardness index, IHI, was calculated by dividing the maximum load, L (in kN), applied to the specimen to the maximum penetration, D (in mm). IHI = L/D. Based on proposed indentation test, rocks are classified as extremely soft rock if the hardness index is less than 8 and extremely hard rock if the hardness index is more than 50. For the tests showing chipping phase, the peak load and penetration were taken at the point of the first chipping. For those tests that did not display any chipping phase, the maximum load and penetration values were taken at the load up to 20 kN or penetration reaching 1 mm. In order to assess the precision and to calculate the average value, two tests were conducted on each specimen at diametrically opposite locations. The indentation hardness index and uniaxial compressive strength were determined for a wide range of rock samples. Each sample was inspected for microscopic defects so that it would provide a test specimen free of fractures or visible planes of weakness. Both the UCS and IHI are indicators of the rock's resistance to elasto-plastic deformation. They are roughly related as UCS = 3.1 * IHI$^{(1.09)}$. It was said that a better correlation between UCS and IHI can be obtained if the IHI is calculated as a tangent value to the load penetration profile rather than the average value. It was further said that standardized indentation testing allows for the characterization of mechanical properties of rock and there is a relationship between the value of the indentation hardness index and the UCS. Further, it was also concluded that the value of the calculated index can be used to classify the hardness of rock and serve as an independent method for assessment of rock strength.

3.2.3 Number of Samples Required for Determining Mechanical Properties of Rocks

Gill et al. (2005), through rigorous statistical method, proposed an algorithm to determine the minimum number of specimens (optimal sample size) required in the laboratory testing for the determination of mechanical properties of rocks. All the basic equations used in the algorithm were based on small sampling theory or exact sampling theory. Optimal sample size may be defined as the smallest number of specimens in a group for given rock and test, which ensures the required precision index for a given confidence interval. Sampling theory is essentially the study of various relationships existing between population parameters (e.g., true mean, variance) and sample (e.g., observed mean, variance) drawn from the population. Investigators assumed that the rock property the engineer is testing is right for his design. In the algorithm, the relationships between the number of specimens in a group, the coefficient of variation obtained after testing the specimen, the targeted precision index and confidence intervals were considered. Confidence limits of the true mean (population mean) using the sample mean (observed mean) is given by $\overline{X} - t_\beta \frac{s}{\sqrt{N-1}} \le \mu \le \overline{X} + t_\beta \frac{s}{\sqrt{N-1}}$ where μ is the population arithmetic mean, tβ represents the confidence coefficients obtained from the student t distribution, β is a parameter determined by the required confidence level. The hypotheses to be satisfied for the estimation of confidence intervals were:

a. Sampling procedure should be such that all population elements have the same probability of being selected;
b. Sampling is done from an infinite population of elements and sample elements are independent;
c. The values of the character measured on sample elements are normally distributed throughout the population.

The precision index is defined as the ratio of upper and lower bounds of the population mean interval as estimated from a group of specimens. It was said that the precision index is a function of the confidence coefficient and of the sample size. If confidence interval is increased, the sample size must be increased to maintain a given precision index. To test the validity of the proposed optimization algorithm, three simulation sets were performed. By attaining a given precision index, it was ensured that the true mean and standard deviation are both within the range defined by the precision index and sample size with a chosen confidence interval of 95%. The simulations were carried out using the Monte Carlo technique. The data given by this technique follows a Gaussian distribution with a mean and standard deviation equal to that of the population. It was concluded that for a given rock and test types, the minimal sample size varies from one case to another, because of which it is impossible to determine a priori the sample size required to obtain a given precision index for a certain confidence level. It was also emphasized that the investigation dealt with the way test data should be considered from a statistical point of view. Choosing what rock properties are to be tested and how the values obtained are used in design modeling are up to the rock engineer's judgment and experience.

3.2.4 Determination of Specific Energy

Teale (1965) analyzed the drilling data from a number of sources from various researchers and proposed a concept of specific energy in rock drilling. When the tool is forced into the rock surface, the solid rock cuts into small fragments of assorted sizes or crushing takes place. In a crushing process, energy/volume relationships are very important. *Specific energy* is defined as the energy required to excavate a unit volume of rock. (The quantity of rock broken is logically measured geometrically, by volume rather than by mass, since it is determined by stress pattern, which is itself geometrical.) Specific energy can be taken as an index of the mechanical efficiency of a rock-working process. The mechanical efficiency of a rock-working process cannot be measured directly, since the minimum amount of work required to excavate a given volume of rock cannot be measured directly. Therefore, the volume of rock actually broken must be provided with an index. The volume of rock broken per unit energy input is the reciprocal of specific energy, and therefore the mechanical efficiency is maximum when specific energy is minimum. In rotary nonpercussive drilling, work is done both by the thrust (F *lb*) and the torque (T *lb.in.*). If rotational speed is N rev/min, the area of the hole or excavation A in^2 and penetration rate u in/min, the total work done in 1 minute is (Fu + 2πNT) *in.lb*. The volume of rock excavated in 1 minute is Au in^3.

$$\text{Therefore, specific energy } e = \left(\frac{F}{A}\right) + \left(\frac{2\pi}{A}\right)\left(\frac{NT}{u}\right) in.lb / in^3$$

$$e = e_{thrust} + e_{rotary}$$

where $e_t = \left(\frac{F}{A}\right) in.lb./in^3$ and $e_r = \left(\frac{2\pi}{A}\right)\left(\frac{NT}{u}\right) in.lb/in^3$

It was derived that the thrust component (F/A) is equivalent to the "mean pressure" exerted by the thrust over the cross-sectional area of the hole. For a given size of excavation, "A" is constant so that "e_t" is directly proportional to "F." For a given value of "A" and "N," "e_r" is proportional to (T/u). Now the torque/penetration rate curves for rotary drilling, over a fairly wide working range, approximate to straight lines through the origin. Thus (T/u), which would be the slope of such a line is also approximately constant. It follows that for given "A" and "N," "e_r" and therefore "e" itself should not vary a great deal over the given working range. This is consistent with the suggested constant value of specific energy at large particle sizes. Another approach for the explanation of "e_r" is to put "p" as the penetration per revolution (p = u/N in./rev.). Then, $e_r = \left(\frac{2\pi}{A}\right)\left(\frac{T}{p}\right) in.lb/in^3$. T is the torque required to remove a layer of rock of depth "p" in one revolution. Since the amount of energy required to break brittle materials like rock is not much affected by the rate at which it is applied, the relationship between "T" and "p" may not be significantly affected by changes in rotation speed. The ratio (T/p) may therefore be a useful index of specific energy. As the thrust increases, the size of particle broken will also increase; the work lost in friction will constitute a rapidly decreasing percentage of the total work done. Together, these effects will contribute to a fall in specific energy. However, this fall will not continue indefinitely; a stage may be reached when the tool is pushed so heavily into the rock that it becomes overloaded and clogs. The reduction in efficiency at this stage will cause the specific energy to rise again until the drill stalls. For a practical drilling tool, then, operating at a fixed rotation speed in a particular rock, the specific energy will be high at low thrusts. Specific energy will fall rapidly as the thrust increases, until it reaches a value beyond which it either will continue to decrease so slowly as to remain virtually constant or will actually start to rise again. The lowest value attained is a measure of maximum mechanical efficiency of the particular tool under specific operating conditions. Minimum specific energy is correlated with crushing strength for different drilling tests such as the roller cutter test, test with tri-cone roller bit (Security M3, Hughes W73), Russian tunneling machine (Protodyakonov scale is used to denote the strength of rocks), small-hole rotary drilling (orthodox two-winged carbide tipped rotary bits with neutral, negative, and positive rakes), and percussive rotary drilling. The rough correlation between specific energy and crushing strength was confirmed while analyzing the experimental data from a number of research sources of the previously mentioned tests. It was said that understanding the way specific energy is governed by the variables such as thrust and rotation speed is necessary if tools are to be designed and used most effectively. However, percussive rotary drilling results showed that specific energy is governed mainly by geometry, the shape of the bit, and penetration per revolution, and not much influenced by rotation speed or the manner in which the energy is applied. It was concluded that rock-working processes should be studied as mechanical systems for breaking rock into fragments. It was also suggested that in this context, the work done per unit volume broken is the basic factor that relates the rock-working processes to physical properties of the rock and against which the behavior of all other variables should be set.

The specific energy of a rock type is the energy required to break a unit volume of rock using a particular breakage method. However, specific energy is highly dependent on the mode of rock breakage, and the size and nature of the equipment used. Reddish and Yasar (1996) developed a new strength index based on specific energy. During the investigation of specific energy it was observed that larger drills have more torque available and will cut rocks more efficiently. After analyzing the drill results for uniaxial compressive strength and Young's modulus in a number of ways, investigators derived a suitable method of determining a reliable strength index from specific energy. The results were plotted as a graph of specific energy against the penetration rate. The gradient of graph and stall penetration rate provide a useful strength index for the tested rock type. It was said that the efficiency of a drill depends on power, thrust, physical size, mode of breakage (rotary or percussive), bit geometry, and bit sharpness. Of these factors, the bit sharpness and available power have most influence on the tests conducted. A rock drilling specific energy is influenced by parameters such as rock strength, rock stiffness, presence of structural discontinuities, abrasivity and hardness of the mineral constituents, nature of rock matrix, and nature of the mineral grain. From the results, surface effects or irregular grain effects are eliminated because the developed index values reflect the drilling of many centimeters of rock. During the analyses of result, rock strength was reflected by penetration rate and also the gradient of the line through a particular rock's result. With the drill having a controlled amount of power, the maximum rate at which it drills is a function of the rock's strength.

3.2.5 Usefulness of Determining Rock Mass Drillability Index

Hoseinie et al. (2008) proposed a new classification system for specifying the rock mass drillabilty index (RD_i). For this purpose, six parameters of the rock mass, including texture and grain size, Mohr's hardness, uniaxial compressive strength (UCS), joint spacing, joint filling (aperture), and joint dipping were selected. In order to quantitatively study the joints' effects on drilling efficiency, rock masses with specific mechanical and physical specification were physically modeled using concrete. For this, joint spacing as well as dipping and filling of joints of rock were simulated in the laboratory by using concrete blocks. These models were drilled using rotary drilling in the mine. The results of investigation on physical modeling as well as field experience indicated that with increase in joint spacing, the penetration rate of drilling will logarithmically increase and, thus, at the same ratio, drilling efficiency will be enhanced. Similarly, by an increase of joint filling size, drilling rate will increase logarithmically. In this case, critical condition occurs when the dimensions of the filling materials approach the flashing particles. The results of physical modeling suggest that with an increase of the angle between the joint surface and direction of the hole, the penetration rate of the drilling system and the rock mass drill-ability will increase linearly. In order to rate the six parameters, it was considered that the properties relating to the rock material (texture and grain size, UCS, Mohr's hardness) are more important than the structural parameters (joint spacing, dipping, and filling) of the rock mass. This is related to the direct contact of the drill bit with the rock material and substantial effect of physical properties of the rock on the penetration rate of drilling system. Thus, in the proposed classification system, the distinctions related to the rock material have been given more importance than joint properties. In the proposed RD_i system, each rock mass is assigned a rating from 7 to 100, with a higher rating corresponding to greater ease of drilling. Based on the RD_i rating, the drilling rate may be classified into five modes; slow, 10%; slow–medium, 25%; medium, 50%; medium–fast, 70%; and fast, 100%. In order to study the practical ability of the suggested classification system (RD_i), a field study was made on the drill-ability of Sungun Copper Mine's rock masses. In this study, Sungun Copper Mine's rock masses were classified using the RD_i as well as Wilbur classification system. Study shows that with increase in RD_i, penetration rate of Down the Hole (DTH) drilling increases exponentially. It was concluded that the RD_i index has been suggested according to uncontrollable factors governed by in situ geological and mechanical conditions to predict the drilling rate qualitatively. However, for developing a comprehensive drill-ability index of rock masses, and quantitatively prediction, the parameters of drilling equipment and effect of drilling methods (rotary, percussion, percussion–rotary) should be incorporated.

From the results of field investigation, Lama and Vutukuri (1978) and Carmichael (1982) tabulated extensive lists of various mechanical properties of sedimentary rocks from different locations around the world. Kwasniewski (1989) listed UCS and porosity data of various sandstones. Jizba (1991) presented mechanical properties of sandstones and shales with a wide range of porosity recovered from different depths in a borehole in Texas. Wong et al. (1997) presented a table of strength and other physical properties of several representative porous sandstones. Bradford et al. (1998) and Horsrud (2001) reported laboratory test results on the North Sea sandstone and shale, respectively.

This chapter has discussed in detail the application of sound level in rock mechanics like determination of rock properties and usefulness of sound level in determining rock properties. For further details, readers are advised to go through the relevant references.

References

Aleotti, L., Poletto, F., Miranda, F., Corubolo, P., Abramo, F., and Craglietto, A. (1999). "Seismic while drilling technology: use and analysis of the drill-bit seismic source in a cross-hole survey." *Geophys. Prospect.*, 47, 25–39.

Anderson, J.A. (1995). *An Introduction To Neural Networks*. Cambridge, MA, The MIT Press.

Asanuma, H., and Niitsuma, H. (1996). "Triaxial seismic measurement while drilling and estimation of subsurface structure." *Int. J. Rock Mech. Min. Sci. Geomech. Abstr.*, 33(7), 307A.

Aydin, A. (2009). "ISRM suggested method for determination of Schmidt hammer rebound hardness: Revised version." *Int. J. Rock Mech. Min. Sci.*, 46, 627–634.

Bradford, I.D.R., Fuller, J., Thompson, P.J., and Walsgrove, T.R. (1998). "Benefits of assessing the solids production risk in a North Sea reservoir using elastoplastic modeling." *Proc., SPE/ISRM Eurock '98*, Trondheim, Norway, 261–269.

Carmichael, R.S. (1982). *Handbook of Physical Properties of Rocks—Vol. II*. Boca Raton, FL, CRC Press.

Gill, D.E., Corthesy, R., and Leite, M.H. (2005). "Determining the minimal number of specimens for laboratory testing of rock properties." *J. Eng. Geol.*, 78(1–2), 29–51.

Haykin, S. (1999). *Neural Networks—A Comprehensive Foundation*. Prentice Hall, Upper Saddle River, New Jersey.

Horsrud, P. (2001). "Estimating mechanical properties of shale from empirical correlations." *SPE Drill Complet*. 16, 68–73.

Hoseinie, S. H., Aghababari, H., and Pourrahimian, Y., (2008); "Development of a new classification system for assessing of rock mass drillabilty index (RDi)." *Int. J. Rock Mech. Min. Sci.*, 45(1), 1–10.

Hsu, K. (1997). "Sonic-while-drilling tool detects over pressured formations." *Oil Gas J.*, 59–67.

Jizba, D. (1991). "Mechanical and Acoustical Properties of Sandstones and Shales." Ph.D. thesis, Stanford University.

Kahraman, S., Korkmazve, S., and Akcay, M. (1996). "The reliability of using Schmidt hammer and point load strength test in assessing uniaxial compressive strength." K.T.U. Dept. Geol. Eng. 30th year Symp. Book, Trabzon, 362–369.

Kawasaki, S., Yoshida, M., Tanimoto, C., and Masuya, T. (2000). "The development of property evaluation method for rock materials based on the simple rebound hardness test: Investigations on the effects of test conditions and fundamental properties." *J. Jpn. Soc. Eng. Geol.*, 41, 230–241.

Kawasaki, S., Tanimoto, C., Koizumi, K., and Ishikawa, M. (2002). "An attempt to estimate mechanical properties of rocks using the Equotip hardness tester." *J. Jpn. Soc. Eng. Geol.*, 43, 244–248.

Kwasniewski, M. (1989). *Rock at Great Depth*. Laws of brittle failure and of B–D transition in sandstones, V. Maury and D. Fourmaintraux, eds., Brookfield, Balkema, state 45–58.

Lama, R.D. and Vutukuri, V.S. (1978). *Handbook on Mechanical Properties of Rocks—Vol. II*. Trans. Tech. Publications, Clausthal, Germany.

McCarroll, D. (1994). The Schmidt hammer as a measure of degree of rock surface weathering and terrain age. In Beck C. editor. *Dating in Exposed and Surface Contexts*. Albuquerque: Univ. New Mexico Press, pp. 29–45.

Miranda, F. (1996). "Impact of the seismic while drilling technique on exploration wells." *Int. J. Rock Mech. Min. Sci. Geomech. Abstr.*, 33(8), 360A.

Proceq, S.A. (1977a). *Operating Instructions Concrete Test Hammer Types N and NR*. Zurich, Switzerland.

Proceq, S.A. (1977b). *Equotip Operations Instructions*. 5th edn. PROCEQ SA Zurich, Switzerland.

Rector, J.W., and Hardage, B.A. (1992). "Radiation pattern and seismic waves generated by a working roller-cone drill bit." *Geophysics*, 57(10), 1319–1333.

Reddish, D.J., and Yasar, E. (1996); "A new portable rock strength index test based on specific energy of drilling." *Int. J. Rock Mech. Min. Sci. Geomech. Abstr.*, 33(5), 543–548.

Schmidt, E. (1951). "A non-destructive concrete tester." *Concrete*, 59, 34–35.

Szlavin, J. (1974). "Relationship between some physical properties of rock determined by laboratory tests." *Int. J. Rock Mech. Min. Sci. Geomech. Abstr.*, 11(2), 57–66.

Szwedzicki, T. (1998). "Technical note—indentation hardness testing of rock." *Int. J. Rock Mech. Min. Sci.*, 36(6), 825–829.

Teale, R. (1965); "The concept of specific energy in rock drilling." *Int. J. Rock Mech. Min. Sci.*, 2(1), 57–73.

Tsuru, T., and Kozawa, T. (1998). "Noise characterization in SWD survey." Soc. Explor. Geophys. Japan, Tokyo, Butsuri-Tansa, *Geophys. Explor.*, 51(1), 45–54.

Ulusay, R., and Hudson, J.A., eds., (2007). "The complete ISRM suggested methods for rock characterization, testing and monitoring: 1974–2006." Compilation arranged by the ISRM Turkish National Group, Ankara, Turkey, 628 p.

Verwaal, W., and Mulder. A. (1993). "Estimating rock strength with the Equotip hardness tester." *Int. J. Rock Mech. Min. Sci. Geomech. Abstr.*, 30, 659–662.

Wong, T.F., David, C., and Zhu, W. (1997). "The transition from brittle faulting to cataclastic flow in porous sandstones: mechanical deformation." *J. Geophys. Res.*, 102, 3009–3025.

Zhang, L. (2005). *Engineering Properties of Rocks*. Elsevier Publications, Amsterdam, 4, 1–290.

4

Application of Acoustic Emission and Equivalent Sound Levels in the Geotechnical Fields

4.1 Introduction

Generally, the process of drilling always produces noise as a by-product. This noise is generated from the bit–rock interface, regardless of the type of bit or material the bit is drilling in (rock, wood, concrete, metal). During the process of drilling, it is important to know the type of rock being drilled. One possible way to determine the type of rock is to analyze the noise produced during drilling by identifying the specific acoustic signature of each drilled rock.

4.2 Application of Acoustic Emission in Geotechnical Engineering

Acoustic emission studies related to rocks were initiated by Obert (1941) and Obert and Duvall (1942) to predict rock bursts in mines. Many authors (Knill et al., 1968; Hardy, 1972; Hardy, 1977; Mogi, 1973; Shamina, 1975; Byerlee, 1978) focused attention on the change in spectral amplitudes of acoustic waves over a wide band of monitoring frequencies with increasing applied stress.

McNally (1990) proposed an exponential relationship between uniaxial compressive strength (UCS) and the sonic log after study of the UCS test results on thousands of core samples with geophysical logs from Bowen Basin. This relationship is widely accepted as a conventional approach called the McNally Method in Australian coal mining for estimation of the UCS from the sonic log. The McNally method is a first-order estimate of rock strength. It has been found that at many mines, a local relationship is required to enable estimation of UCS with sufficient accuracy. To rectify this problem, the same investigator suggested a number of mathematical equations/expressions for weak and strong strata. The German Creek Mine derived its own local formula (Ward 1998).

Hatherly (2002) proposed an alternative approach to estimate the UCS of elastic rocks from geophysical borehole logs. The investigator first determined the composition of the rocks and then correlated this to the UCS. Behind this proposition was the idea that compositional parameters would have a major influence on rock strength. Schon (1996) found that there is either a linear, polynomial, or logarithmic relationship between seismic velocity and UCS.

During the process of rotary drilling, noise is produced from the drilling machine engine, contact between the drill bit and the rock, and from the cooling water. Zborovjan (2001) showed that relevant rock acoustic signature can be found between 5000 and 8000 Hz. It was said that the lower half of the signal spectrum is mainly due to the noise produced from the drilling engine and cooling water.

Zborovjan et al. (2003) discussed acoustic identification of rocks during the drilling process. An attempt was made by the investigators to identify the specific acoustic signature of each drilled rock type and rock class. Software was developed for acoustic rock identification based on hidden Markov models (Rasmussen 2000). The developed software included the real experimental results to identify the specific acoustic signature of each drilled rock type/rock class. It was said that the proposed software gives satisfying results in rock class acoustic identification.

Miklusova et al. (2006) developed an experimental setup to simulate rock disintegration by rotary drilling in the laboratory. The setup provides rotary drilling of rock samples using small diameter drilling tools. The drilling stand of the experimental setup was equipped with monitoring systems providing the scanning, recording/computation of input and output variables of disintegration process, such as thrust, revolutions, advance rate, specific disintegration energy, etc. The experimental setup was also provided with a microphone at a specific point in acoustic space, which registered an acoustic signal arising during the drilling process. The acoustic signal was processed with the Fourier transformation decomposing general in-harmonic periodical action into harmonic compounds (i.e., variety of additional overtones into a single distinct sound). The analysis of noise signals as an acoustic behavior of rock drilling process was discussed. It was also emphasized that acoustic signal analysis has the potential to be used for control of rock disintegration process.

Futo et al. (2003) discussed on the possibility of rock drilling optimization through acoustic signals. During rock disintegration process, the thrust on the drilling tool, revolutions, drilling speed, and disintegration power were monitored. The investigators designed a system for noise monitoring during the rock disintegration process. The experimental results for the optimization of disintegration process with the use of acoustic signals were also carried out. An algorithm was developed for determination of optimal parameters. As per this algorithm, the equivalent sound levels of a drill machine during idle, as well as during the rock disintegration process, should be determined. The next step should be to find the equivalent sound level by changing the drilling parameters. It was said that if the difference of the equivalent sound level between idle condition of drill machine and during rock disintegration process increases, then the input drilling conditions needs to be changed. If the difference decreases, determination of the equivalent sound level during rock disintegration should be continued. The finding of the equivalent sound level by changing of drilling parameters should be repeated until the required accuracy is achieved. For the development of optimal control algorithm, the results of research in the field of acoustic signals for optimal control of drilling machine were used. The algorithm was verified in laboratory conditions. It was concluded that the equivalent level for representative frequency depends on the rock types, and therefore equivalent sound levels can be used for identification of the rock type.

Gradl et al. (2008) claimed that bit characteristics can be determined using only acoustical data, i.e., noise of a bit. The resulting noise of the bit/rock interaction while drilling was recorded using a standard microphone literally pointed at the bit and analyzed. Using the fast Fourier transform, frequency spectra were extracted from the recorded data. Investigators could detect major differences in the frequency characteristics during the analysis of the roller cone bit, the PDC, and the natural diamond bit. For drag bits, the frequency characteristics could be related to the bit's design. The frequency spectra of the roller cone bit with a general high amplitude level are relatively evenly distributed over the whole frequency spectrum. Although the objective of the research was to determine whether drill bits used in petroleum engineering applications have unique frequency characteristics, and if found to be true, then whether these frequency characteristics can be linked to bit design parameters. Potential applications of this research include detecting and diagnosing bit problems (e.g., broken teeth, bit balling) in real time using acoustic data.

Williams and Hagan (2006) investigated whether changes in rock cutting conditions are reflected in the nature of acoustic signals generated in rock. A test facility comprising of a modified Invicta 6M linear rock cutting machine with a triaxial force dynamometer, 12.5-mm-wide tungsten carbide cutting bits, acoustic transducer (accelerometer, amplifier, two 8-channel signal conditioning boards, two analogue-to-digital cards), and a data acquisition system was established. The study involved a series of 12 tests that examined: cutting depth; state of wear of the cutting tool; distance between transducer and cutting groove; and location of the transducer with respect to the cutting direction. During the study, changes between new and worn-out cutter picks, depth of cut, and attenuation of the acoustic signal with distance were observed. The acoustic emission was found to vary with time during rock cutting, and there seemed to be correlation between the rise in cutting force and the levels of acoustic emission. It was also found that changes in rock cutting conditions appeared to have some measurable impact on the nature of the acoustic emission.

The acoustic emission rate (AE) and the total number of events are the usual measurement parameters for determining the intensity of AE. The AE system is designed to detect and process the emissions from yielding, crack growth, corrosion, and other sources of AE. Thus, a certain amount of work is done. This work is converted to potential energy of AE. The energy is an integral part of the wave-packetsful wave envelope. This can be expressed as

$$E = \int_{t_1}^{t_2} \left[V(t - t_1)e^{-(t-t_1)}\beta \sin \omega_0^{(t-t_1)} \right]^2 dt$$

where V = maximum burst voltage, β = transducer damping factor, and ω_0 = band pass broad center frequency.

Jung et al. (1994), conducted experiments on rock core specimens (both hard and soft rock types). These include Orofino gneiss, white cloud granite, Clark fork quartzite, decite, Boise sandstone, mudstone, quartzite, siltite, and argillite from Lucky Friday Mine. During indentation, the cores were confined with a steel ring. A 3-mm-diameter flat-bottomed circular punch was used for indentation. The controlled piston rate was 8.5×10^{-3} mm/sec. Load displacement, AE RMS time, AE number time, and AE signal time curves were recorded and analyzed. During indentation, AE activities increased with increasing stress level and rose sharply just as the critical load was reached and chipping occurred. The AE rate, total event number, peak value of the AE mean amplitude, total AE energy, etc., to a large extent depend on rock hardness, which is closely associated with its mineral composition, structure and texture, and physical properties. In the study, indentation tests have indicated that an obvious correlation exists between indentation hardness and AE parameters. The peak RMS related well to indentation hardness, which was defined as peak indentation load per unit contact area. The more energy necessary to indent, the higher was the resulting AE. For each indentation test, the indentation energy was measured on the load-displacement curve. Then, the average indentation hardness (H_a), average indentation energy (E_a), and average integrated RMS (RMS_a) were calculated for each kind of rock. It was seen that H_a, E_a, and RMS_a were fairly identical for nine rocks. This result confirmed that the integrated RMS of AE events possessed the energy intention. The RMS of AE events is the average signal level or the mean amplitude of AE. It reflects the average signal strength. At the strength of the material, it was found that the sample showed a pronounced maximum of the average AE amplitude near the yield point. The yield point maximum of the mean amplitude was corresponding to the strong dislocation. In rock indentation, the obvious peak RMS was generated immediately after the chip formation. The greater the indentation hardness of rock, the more energy was needed to be accumulated to create chips, and greater AE was released after chipping. Rock indentation tests indicated that there was a certain relationship between the indentation hardness of rock and the AE parameters, such as accumulated number of events, peak RMS, and integrated RMS. It was concluded that, with the above-mentioned parameters, the correlation of RMS of AE events with indentation hardness is pronounced and the integrated RMS is the best index to correlate with the indentation hardness of rock. It represents that the sum of amplitudes of AE varies during all the indentation process and reflects the size of AE energy released in the time period of indentation.

Vardhan and Murthy (2007) carried out an experimental investigation of jackhammer drill noise with drilling in rocks of different compressive strength. The authors investigated the influence on sound level due to drilling in rocks of varying physical properties like compressive strength and abrasivity using a fabricated jackhammer drill set up. Five different rock samples were used to find out the sound level at different positions of measurements like near the drill bit, drill rod, exhaust, and at operator's position of the fabricated jackhammer drill set up.

An attempt was also made by Vardhan et al. (2009) to investigate the usefulness of sound level in determining rock or rock mass properties like compressive strength using jackhammer drill on laboratory scale by fabricating a jackhammer drill setup wherein the thrust applied can be varied while drilling vertical holes (Figure 4.1). It was concluded that sound level produced during drilling will vary with the rock properties. However, investigators suggested for further work in this direction to confirm the variations and possible prediction of rock properties by conducting more detailed experiments with a large variety of rock samples.

4.3 Regression and ANN Techniques in Geotechnical Fields

In recent years there has been a steady increase in the successful and potential application of regression and neural network modeling in many different areas of geosciences.

4.3.1 Applications of Regression Techniques

Sachapazis (1990) collected and tested 29 different types of carbonate rocks from various sites in Greece and four from disused quarries in England (Longhoughton and Mootlaw). The tests included the determination of Schmidt hammer rebound hardness number, tangent Young's modulus, and uniaxial compressive strength. Finally, these parameters were correlated and regression equations of high practical value were established among the properties noted, all with high coefficients of determination (R^2).

Kim and Gao (1995) proposed a statistical approach for the calculation of the mechanical properties of rock mass. It was said that the approach accounts for the uncertainty due to the variability of the rock material

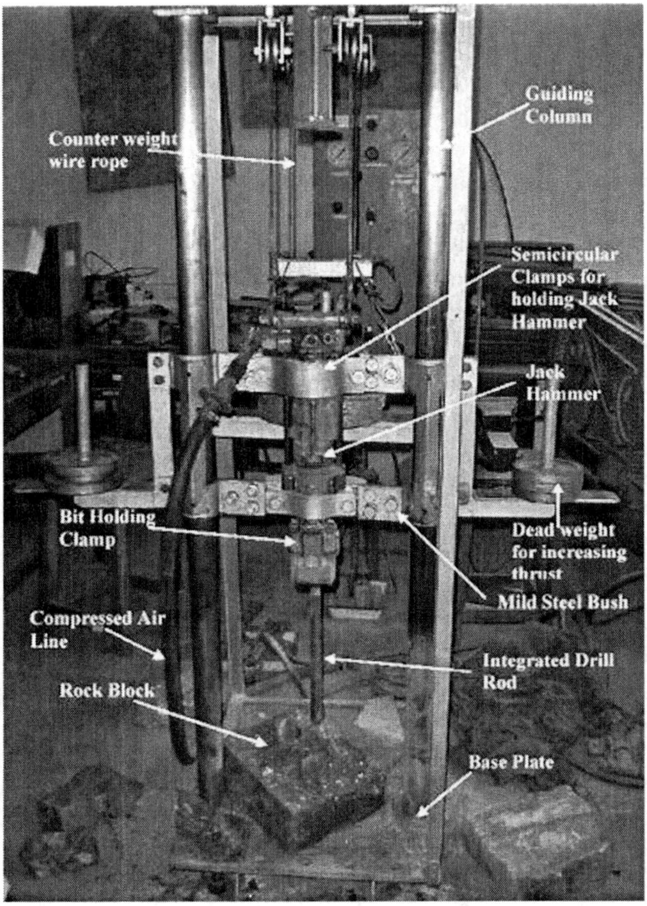

FIGURE 4.1
Pneumatic drill setup for drilling vertical holes in rock blocks.

properties and the pattern of the discontinuities in rock mass. All parameters describing the rock mass properties were considered random variables instead of a constant. Probability distributions of the deformation modulus and the compressive strength of a rock mass were obtained by applying Monte Carlo simulation. Using extreme value statistics, the input data used for the simulation along with the simulation results were analyzed. It was said that the third type asymptotic distribution of the smallest value is the general statistical representation of mechanical properties of rocks.

Katz et al. (2000) used Schmidt hammer rebound readings, of seven rock types (Maresha chalk, Cordoba cream limestone, Berea sandstone, Indiana limestone, Carrara marble, Gevanim syenite, and Mt. Scott granite) to develop new empirical correlations with laboratory measured values of Young's modulus, uniaxial compressive strength and density. It was said that these developed equations can be used to estimate the relevant mechanical properties in the field and laboratory. However, some precautions should be observed, which include the following: The tested rock is well cemented and apparently elastic. Rocks that tend to disintegrate under hammer impact or samples that crack under the impacts cannot be properly tested. Hammer measurements should be conducted on smooth surfaces especially for fieldwork. Fractured blocks can be measured if the intact part of the block weighs a few tens of kilograms or more.

Kahraman (2001) evaluated the correlations using least squares regression between the uniaxial compressive strength (UCS) values of 48 different rocks and the corresponding results of point load, Schmidt hammer, sound velocity, and impact strength tests. The variability of test results for each test and each rock type was evaluated by calculating the coefficient of variation. It was found that, strong linear relations exist between the point load strength index values and UCS values for the coal measure rocks and other associated rocks. The Schmidt hammer and sound velocity tests exhibit significant nonlinear correlations with the compressive strength of rock. In the sound velocity test, the data points were scattered at higher strength values. There was no clear relation between the impact strength values and compressive strength values for coal measure rocks. A weak nonlinear correlation was found between the impact strength values and the compressive strength values for the other rocks. It was concluded that, except the impact strength, all other test methods evaluated in the study provide reliable estimation of the compressive strength of rock.

Rock indentation is the basic process in drilling and mechanical excavation. The mechanics of rock failure and chip formation beneath a drill bit is a complicated function of the mechanical and physical properties of the rock, the state of the stress at the bottom of a borehole, the geometrical design and shape of the bit teeth, and the type and flow rate of drilling fluid. Kahraman et al. (2000) developed a model for the prediction of rotary blasthole drills using drillability index from the force–indentation curves. The relationships between drillabilty index and rock properties were studied, and two different bit teeth (conical and spherical) were also compared. Twenty-two rock samples from various drilling locations were collected, and physical and mechanical properties were determined. Bit type and diameter, hole length, feed pressure, rotational pressure, air pressure, and net drilling time were recorded during the performance studies. Then average penetration rates were calculated. In the laboratory, using the servo-controlled stiff-testing machine, indentation tests were executed. For indentation a conical bit tooth of 60° apex angle and a spherical bit tooth with a diameter of 14.4 mm were used. In the tests, the load was measured by a load cell of 0.22 MN capacity, and the displacement (indentation) was measured by a transducer (LVDT) of 0.25 mm/V. The test duration was 3 min with 2 mm/min loading rate and 3 mm penetration depth. During the study of force–indentation curves, obtained from indentation tests, linear curves were obtained for plastic rocks that were only crushed and penetrated by an indenter. Nonlinear and discontinuous curves were obtained for brittle rocks that were cracked with chips formed. Stepped curves were seen on the force-indentation curves of very soft formations (σ_C < 25 MPa), except the sandstone formation. Drillabilty index was described as, for the force-indentation curves giving chip formation, the average slope of the tangent plotted for 50% value of the chipping force. For the curves not giving chip formation, the slope of the force-indentation curve was taken at the half of the maximum penetration value. An empirical equation was formulated for drilling rate prediction, which is a function of rotational speed, penetration rate, bit thrust, bit diameter, drillabilty index, and proportionality constant "k"; in situ tests were carried out similar to laboratory tests. The measured parameters were inserted into a predictor equation, and the relationship between the proportionality constant, drillabilty index, and physical and mechanical properties were worked out. Finally, an equation for the penetration rate was formulated, which is a function of a rock's mechanical properties, rotational speed, bit thrust force, and bit diameter. It was shown that there was a linear relationship between the uniaxial compressive strength and drillabilty index. However, it was found that the conical bit tooth follows a more sloped line than the spherical bit tooth. Similar result was found for tensile strength and point load index versus drillabilty index. The product of Schmidt hammer rebound number and density with drillabilty index increases first rapidly and then gradually. There was a nonlinear relationship between impact strength, P-wave velocity, elastic modulus, and density versus drillability index. During the comparison of conical and spherical bit tooth, it was stated that when the rock properties are constant, the drillabilty index of the spherical bit tooth is greater than the drillability index of the conical bit tooth. Also for a given penetration depth, the penetration force for the spherical bit tooth was found to be greater than the penetration force for the conical bit tooth in the same rock. It was concluded that the model proposed is valid for rock types having compressive strength over 40 MPa, and it emphasizes that penetration rates of blast hole drills may be predicted accurately from drill bit diameter, rotational speed, thrust force, and one of the rock properties.

Altindag (2002) investigated the relationships between brittleness concepts and different rock properties (drillability index, point load, elastic modulus) with reference to the data from the experimental variables of Bilgin et al. (1993) and Kahraman et al. (2000), using regression analysis. It was found that there exists a significant correlation between brittleness and the drillability index.

Kahraman (2002) investigated the relationship between the direct P-wave velocity value and the indirect P-wave velocity values obtained from laboratory measurements.

A number of studies on the investigation of usefulness of the Schmidt hammer test on different rock types have been made by Deere and Miller (1966), Aufmuth (1973), Beverly et al. (1979), Kidybinski (1980), Singh et al. (1983),

Sheorey et al. (1984), Haramy and DeMarco (1985), Ghose and Chakraborti (1986), O'Rourke (1989), Cargill and Shakoor (1990), Kahraman et al. (1996), Tugrul and Zarif (1999), and Yilmaz and Sendir (2002). Various empirical equations have been proposed for estimating UCS and Young's modulus of rock from Schmidt hammer rebound number, which is given in Table 4.1.

Yasar and Erdogan (2004a) studied the statistical relationships between Schmidt hammer rebound number and Shore scleroscope hardness values with physico-mechanical properties such as unit volume weight, porosity, strength to blow, strength to bending, average abrasion strength of Ceyhan limestone, Barbados marble, antique cream limestone, Osmaniye marble, hardness sandstone, and Toprakale basalt rocks. It was concluded by comparing the results of different regression equations with the laboratory measured values, that physico-mechanical properties can be estimated using hardness methods. It was also shown that the porosity increases with decreasing hardness values and strength of the rocks.

Yasar and Erdogan (2004b) developed statistical relations using the least square regression method to predict physico-mechanical properties (UCS,

TABLE 4.1

Correlation between Schmidt Hammer Hardness with Uniaxial Compressive Strength (UCS) and Young's Modulus (E)

Equation	R^2	Researcher
$UCS = 6.9 \times 10^{[0.0087\rho N + 0.16]}$	0.94	Deere and Miller (1966)
$E = 6.95\rho^2 N - 1.14 \times 10^6$	0.88	
$UCS = 6.9 \times 10^{[1.348 \log(\rho N) - 1.325]}$	—	Aufmuth (1973)
$E = 6.9 \times 10^{[1.06 \log(\rho N) + 1.86]}$	—	
$UCS = 12.74\ e^{(0.185\rho N)}$	—	Beverly et al. (1979)
$E = 192(N\rho^2)12710$	—	
$UCS = 0.447\ e^{[0.045(N + 3.5) + \rho]}$	—	Kidybinski (1980)
$UCS = 2N$	0.72	Singh et al. (1983)
$UCS = 0.4N - 3.6$	0.94	Sheorey et al. (1984)
$UCS = 0.994N - 0.383$	0.70	Haramy and DeMarco (1985)
$UCS = 0.88N - 12.11$	0.87	Ghose and Chakraborti (1986)
$UCS = 702N - 1104$	0.77	O' Rourke (1989)
$UCS = 4.3 \times 10^{-2}(N\rho) + 1.2$ *for sandstones*	—	Cargill and Shakoor (1990)
$UCS = 1.8 \times 10^{-2}(N\rho) + 2.9$ *for carbonates*	—	
$UCS = 4.184N - 65.792$	0.93	Sachpazis (1990)
$UCS = 4.5 \times 10^{-4}(N\rho)^{2.46}$	0.93	Kahraman et al. (1996)
$UCS = 8.36N - 416$	0.87	Tugrul and Zarif (1999)
$UCS = 2.208\ e^{0.067N}$	0.96	Katz et al. (2000)
$E = 0.00013N^{3.09074}$	0.99	
$UCS = e^{(0.818 + 0.059N)}$	0.98	Yilmaz and Sendir (2002)
$E = e^{(1.146 + 0.054N)}$	0.91	

Note: UCS = uniaxial compressive strength (MPa), E = Young's modulus (GPa), N = Schmidt hammer rebound number, ρ = rock density (g/cc).

Young's modulus, and density) of carbonate rocks (dolomite, marble, and limestone) using sound velocity (SV) index value. P&S wave velocity tests were conducted to develop the SV index value. To eliminate any anisotropic effects of the rocks on the measurements, rock types having no bedding planes were selected during sampling. It was concluded that the above said physico-mechanical properties can be estimated from SV values by using developed mathematical equations.

Singh and Rao (2005) conducted a large number of uniaxial compressive strength (UCS) tests on specimens of jointed block mass having various combinations of orientations and different levels of interlocking of joints. Four dominating modes of failure such as splitting of intact material of the elemental blocks, shearing of intact block material, rotating of the blocks, and sliding along the crucial joints were observed. The results obtained on strength and tangent (elastic) modules were used to arrive at the relations linking the properties of the jointed mass with those of the intact rock. These properties depend on the failure mode, which in turn depends on the joint configuration. The validity of the expressions suggested was verified by applying them to the field data obtained from several dam sites from northern India. It was concluded that there was a strong correlation between ultimate strength and the tangent modulus values of the block mass tested in the laboratory for a specific failure mode when plotted on the Dure–Miller classification chart, and it followed an empirical straight line.

Karakus et al. (2005) used multiple regression modeling technique to predict elastic properties of intact rocks from index tests.

Chang et al. (2006) summarized 31 empirical equations that relate unconfined compressive strength and internal friction angle of sedimentary rocks (sandstone, shale, and limestone) to physical properties (such as velocity, modulus, and porosity). From the literature, investigators compiled the ability of these equations to fit laboratory measured strength and physical property data. It was reported that some equations work reasonably well (for example, strength–porosity relationships for sandstone and shale). However, rock strength variations with individual physical property measurements scatter considerably, indicating that most of the empirical equations are not sufficiently generic to fit all the data published on rock strength and physical properties.

Kahraman and Alber (2006) correlated the electrical resistivity values with experimentally measured physico-mechanical properties such as uniaxial compressive strength, elastic modulus, point load strength, Schmidt hammer value, *P*-wave velocity, density and porosity values, using least squares regression method. The derived equations were checked by t and F tests. It was said that UCS and elastic modulus increases with increasing electrical resistivity. Strong logarithmic relationships were found between UCS and resistivity and between resistivity and elastic modulus. It was found that density linearly increases with increasing resistivity. There was an inverse relation between porosity and electrical resistivity. It was concluded that electrical resistivity can be used as a representative measure of rock properties, particularly for characterizing rocks for which regularly shaped specimen are difficult to obtain.

Drilling performance based on the specifications of a drill system, properties of a drill system, and the properties of rock to be drilled can be predicted by drillability. Drillability studies from various researchers indicate that the drilling rate and energy requirement of a diamond drill depend on controllable factors such as drill bit properties (e.g., bit type, bit diameter, diamond quality, size, and geometry) and operational parameters (e.g., weight on bit, rotational speed, and circulating fluid) and uncontrollable factors such as formation properties (rock properties and geological conditions). Huang and Wang (1997) studied the mechanics of diamond coring in rocks. In the laboratory, based on the penetration rate, applied torque, and specific kerfing energy (in case of coring, the volume of rock material removed includes the volume of core and the volume of kerf; the energy required to remove unit volume of kerf is called *specific kerfing energy*), the drillabilty of surface set drill bit was assessed. A modified clausing drill press was used for the study. The drill system includes three load cells mounted to the bottom plate of drill press and connected to a joint box. Weight on a bit was measured by totaling the load measured by each individual load cell. A tachometer with two transducers (placed between the drill spindle and the core barrel) were used to measure the actual rotational speed of the bit and the applied torque. With the long stroke LVDT, the travel of the bit from the rock surface to the cutting face of the hole was measured. The penetration rate was taken as the derivative of the penetration depth with respect to time. From the four transducers, the acquired measurements were transmitted to data acquisition system for processing. Based on an ANOVA test, regression analysis procedure, the mathematical expressions of the relationships among the influential factors were derived. It was concluded that among the three parameters investigated (weight-on bit, rotational speed, and rock type), the weight on bit was most influential factor. The effect of rotational speed of the bit only on the penetration rate (not other measurements) was found to be significant. Penetration rate and applied torque of the drilling increased exponentially as the weight on the bit increased. However, the specific kerfing energy decreased as weight on the bit increased. An exponential relationship between the penetration rates and the uniaxial compressive strengths of rock samples was noted.

Faisal et al. (2007) performed linear statistical analyses to establish the relationships between engineering properties (such as strength, modulus of elasticity, and Poisson's ratio) of the intact rock (dolomite, dolomitic limestone, and shale rocks) and the different types of hardness (Schmidt, shore scleroscope, abrasion, and total hardness). The results showed that there were good relationships between the engineering properties of the intact rock and its hardness.

Yavuz et al. (2008) investigated the abrasion behavior of selected carbonate rocks (limestone, marble, and travertine) by means of three-body abrasion testing using Bohme abrasion testing equipment. Al_2O_3 particles were used as an abrasive material. Dependence of abrasion rate on physico-mechanical properties of rocks (bulk density, effective porosity, P-wave velocity, Schmidt rebound hardness, compressive strength, tensile strength) were investigated through least squares regression analysis. It was concluded that, according to the statistical judgment (i.e., depending on the values of correlation coefficient R^2), hardness, tensile strength, compressive strength, and density of rock could adequately estimate the abrasion rate of rocks, whereas rough estimates could be made with porosity and *P*-wave velocity of rocks. It was said that limestone and marbles tested were more abrasion resistant than the travertine tested. Also, more abrasion-resistant rocks are likely to have high bulk density, compressive strength, tensile strength, hardness, and low porosity.

Kilic and Teymen (2008) determined the statistical equations for estimating the mechanical properties of rocks using nondestructive and indirect test methods. Nineteen different rock types were tested to obtain the relationship between Shore hardness, point load index, sound velocity, Schmidt hardness, porosity, uniaxial compressive strength, indirect tensile strength, and abrasion resistance. Results of regression analysis showed satisfactory correlations.

Yagiz (2008) studied the influence of rock mass properties on tunnel boring machine (TBM) performance and developed a new empirical equation for estimation of the TBM performance. The prediction of TBM performance requires the estimation of penetration rate (ROP), the ratio of excavated distance to the operating time during continuous excavation phase, and advanced rate (AR), the ratio of both mined and supported actual distance to the total time. Rock samples composed of medium grade metamorphic rocks as well as igneous rocks, including gneiss, granite amphibolite, schist, and rhyodacite dykes, were collected from the tunnel site. UCS and Brazilian tensile strength (BTS) were measured according to the procedure suggested by American Society for Testing Materials (ASTM). The punch penetration test was used for investigation of rock brittleness and toughness. The punch penetration test apparatus included a stiff machine with hydraulic ram that presses a tungsten carbide indenter into a saw-cut surface of the sample. While performing the test, the displacement of the indenter into the sample and the load on the indenter were monitored and, consequently, a load versus penetration graph was made. On this graph, the ratio of maximum load (in kN) applied to the specimen to the corresponding displacement (in mm), named the peak slope index (PSI) was used to quantify the rock brittleness and toughness. Rock mass behavior and machine advancement mainly depends on type and density of discontinuities. The alpha angle (angle between tunnel axis and the planes of weakness) was used to quantify the influence of discontinuity properties on TBM performance. An automatic data logging and recording system was fitted to the TBM, in order to monitor the performance of the machine in the field. As a result of using logged data,

TBM penetration rate, total gross thrust, cutter load, net thrust, and torque were computed. In order to perform the statistical analyses for predicting the ROP, the database that is composed of UCS, BTS, and brittleness of intact rock; the alpha angle and distance between planes of weakness (DPW) of rock mass; and also actual measured ROP in the filed were established. Then, one of the commercial software packages (SPSS) was used to perform the forward stepwise regression analyses and thus the influence of each of the rock properties on the ROP was investigated. The significance of the coefficient of correlation was determined by t-test, whereas significance of regression was tested by conducting the F-test. A new performance predictive equation was empirically obtained as a function of measured engineering rock properties: the alpha angle, DPW, PSI, and UCS of rock. In the developed equation, the most effective parameters were the alpha angle, DPW, and PSI of the rock. Highest ROP was found when the alpha angle varied from 50° to 65°. The highest ROP was achieved when DPW varied from 20 to 40 cm. There was a linear relationship between PSI and ROP. Further, ROP increased with PSI. There was no significant impact of UCS and BTS of rock on ROP, so it was excluded from the predictive equation. It was concluded that, even though the study results are reliable for estimating the ROP, the introduced predictive equation should be used with care.

4.3.2 Applications of Artificial Neural Network Techniques (ANN)

Because its multidisciplinary nature and effectiveness, ANN is becoming popular among the researchers, designers, and planners. Therefore, ANN has been successfully used in industrial areas as well. Many researchers used artificial neural networks (ANN) in the geotechnical engineering and other disciplines (Romeo et al., 1995; Miller et al., 1995; Fletcher et al., 1995; Vukelic and Miranda, 1996; Lee and Lee, 1996; Lai and Sera, 1997; Yuanyou et al., 1997; Yang and Zhang, 1998; Alvarez and Babuska, 1999; Alvarez et al., 2000; Finol et al., 2001; Hurtado et al., 2001; Rafiq et al., 2001; Gokceoglu, 2002; Bai et al., 2003; Lee et al., 2003; Basma and Kallas, 2004; Sonmez et al., 2004; Benardos and Kaliampakos, 2004; Kahraman et al., 2006; Sinha and Wang, 2007; Das and Basudhar, 2008; Kayadelen, 2008; Tiryaki, 2008a; Kahraman et al., 2009; Gunaydin, 2009; Taskiran, 2010).

Moulenkamp and Grima (1999) investigated the possibility of predicting UCS by ANN from rock hardness value obtained using Equotip hardness tester and other intact rock properties such as porosity, density, grain size, and rock type information. A dataset containing 194 rock sample records, ranging from weak sandstones to very strong granodiorites, was used to train the network with the Levenberg–Marquardt training algorithm. Two sets, each containing 17 rock samples, were used to validate the generalization and prediction capabilities of the network. The study indicated that ANN technology is more powerful than conventional statistical techniques in predicting UCS from intact rock properties.

Studies of Singh et al. (2001) in developing predictive models for tensile strength, UCS, and axial point load strength from the intrinsic rock properties revealed that using ANN in building a model was more accurate than using conventional statistical techniques.

Sonmez et al. (2006) constructed an artificial neural network model to predict the elastic modulus of the intact rock, by using a multi-input parameter approach. A rock mass rating (RMR) based empirical equation was also proposed for the determination of the deformation modulus of rock masses.

Zorlu et al. (2008) investigated relationships between uniaxial compressive strength and petrographical properties such as quartz content, packing density, and concavo–convex type grain contact of sandstones. A general prediction model using ANN technique was also developed. Also, a routine for the selection of the best prediction model was proposed in the study. The constructed models were checked by using various prediction performance indices and were found suitable for practical use.

Tiryaki (2008b) applied artificial neural networks (ANN) and regression trees technique along with the principal components and factor analyses on the predictions of UCS and E of intact rocks from their other properties such as NCB cone indenter hardness, dry density, and Shore scleroscope hardness. It was reported by evaluation of the graphical and numerical measures of the goodness of the fit, that ANN models were more acceptable for predicting UCS and E than multiple nonlinear regression models. However, regression tree technique has appeared to be the best in building predictive models for UCS and E, among other methods.

Sarkar et al. (2010) reported the use of three-layer feed-forward back-propagation artificial neural network to predict deformation properties of coal measure rocks using dynamic wave velocity, point load index, slake durability index, and density. The results confirmed the applicability of this method. It was reported that this approach can save time and cost. Further research for the development of this method for predicting rock strength parameters were suggested.

Yagiz et al. (2011) developed artificial neural networks and nonlinear regression techniques to assess the influence of slake durability cycles on the prediction of uniaxial compressive strength and modulus of elasticity for carbonate rocks.

Based on this discussion, it appears that rock properties can be determined in the laboratory by various methods which are accurate enough. Also, the applications show that neural network model has superiority in solving problems in which many parameters influence the process and results, even when the process and results are not fully understood. The application of the ANN model is an effort to comprehensively understand the modeling capability of ANN for these kinds of problems, which will really help in better prediction with the available experimental data.

References

Altindag, R. (2002). "The evaluation of rock brittleness concept on rotary blast hole drills." *J. South Afr. Ins. Min. and Metall.*, January/February 2002, 61–66.

Alvarez, G.M., and Babuska, R. (1999). "Fuzzy model for the prediction of unconfined compressive strength of rock samples." *Int. J. Rock Mech. Min. Sci.*, 36, 339–349.

Alvarez, G.M., Bruines, P.A., and Verhoef, P.N.W. (2000). "Modeling tunnel boring machine performance by neuro-fuzzy methods." *Tunnelling Underground Space Technol.*, 15(3), 259–269.

Aufmuth, R.E. (1973). "A systematic determination of engineering criteria for rocks." *Bull. Int. Assoc. Eng. Geol.*, 11, 235–245.

Bai, J., Wild, S., Ware, J.A., and Sabir, B.B. (2003). "Using neural networks to predict workability of concrete incorporating metakaolin and fly ash." *Adv. Eng. Softw.*, 34(11–12), 663–669.

Basma, A.A., and Kallas, N. (2004). "Modeling soil collapse by artificial neural networks." *Geotech. Geol. Eng.*, 22, 427–438.

Benardos, A.G., and Kaliampakos, D.C. (2004). "Modelling TBM performance with artificial neural networks." *Tunnelling Underground Space Technol.*, 19, 597–605.

Beverly, B.E., Schoenwolf, D.A., and Brierly, G.S. (1979). "Correlations of rock index values with engineering properties and the classification of intact rock." FHWA, Washington, D.C.

Bilgin, N., Eskikaya, S., and Dincer, T. (1993). "The performance analysis of large diameter blast hole rotary drills." *J. Min. Mech. Autom.*, 129–135.

Byerlee, J. (1978). "A review of rock mechanics studies in the United States pertinent to earthquake prediction." *Pure Appl. Geophys.*, 116, 586–602.

Cargill, J.S., and Shakoor, A. (1990). "Evaluation of empirical methods for measuring the uniaxial compressive strength." *Int. J. Rock. Mech. Min. Sci.*, 27, 495–503.

Chang, C., Zoback, M.D., and Khaksar, A. (2006). "Empirical relations between rock strength and physical properties in sedimentary rocks." *J. Pet. Sci. Eng.*, 51(3–4), 223–237.

Das, S.K., and Basudhar, P. (2008). "Prediction of residual friction angle of clay artificial neural network." *Eng. Geol.*, 100, 142–145.

Deere, D.U., and Miller, R.P. (1966). "Engineering classification and index properties for intact rocks." *Tech. Report. Air Force Weapons Lab.*, New Mexico, No., AFNL-TR, Kirtland, 65–116.

Faisal, I.S., Edward, J.C. and Omar, H.A. (2007). "Estimation of rock engineering properties using hardness tests." *Eng. Geol.*, 90, 138–147.

Finol, J., Guo, Y.K., and Jing, X.D. (2001). "A rule-based fuzzy model for the prediction of petrophysical rock parameters." *J. Pet. Sci. Eng.*, 29, 97–113.

Fletcher, P., Coveney, P.V., Hughes, T.L., and Methven, C.M. (1995). "Predicting quality and performance of oilfield cements with artificial neural networks and FTIR spectroscopy. *J. Pet. Technol.*, 47(2), 129–130.

Futo, J., Usalova, L., and Ivanicova, L. (2003). "Optimization of rock disintegration using the acoustic signal." *J. Acta Montanistica Slovaca*, 8(4), 174–175.

Ghose, A.K., and Chakraborti, S. (1986). "Empirical strength indices of Indian coals an investigation." *Proc. 27th US Symp. on Rock Mech.*, Balkema, Rotterdam, 59–61.

Gokceoglu, C. (2002). "A fuzzy triangular chart to predict the uniaxial compressive strength of the Ankara Agglomerates from their petrographic composition." *Eng. Geol.*, 66, 39–51.

Gradl, C., Eustes, A.W., and Thonhauser, G. (2008). "An Analysis of Noise Characteristics of Drill Bits." *Society of Petroleum Engineers (SPE), Annual Technical Conference*, Denver, Colorado.

Gunaydin, O. (2009). "Estimation of compaction parameters by using statistical analyses and artificial neural networks." *Environ. Geol.*, 57, 203–215.

Haramy, K.Y., and DeMarco, M.J. (1985). "Use of Schmidt hammer for rock and coal testing." *Proc., 26th US Symp. on Rock Mech.*, Rapid City, 26–28 June, 549–555.

Hardy, H.R., Jr. (1972). "Application of acoustic emission technique to rock mechanics research." *Acoustic Emission*, ASTM STP 505, 41–83.

Hardy, H.R. Jr. (1977). "Emergence of acoustic emission, microseismic activity as a tool in geomechanics." *Proc. First Conf. Acoustic Emission/Micro Seismic Activity in Geologic Structures and Materials*, Trans. Tech. Publications. 13–31.

Hatherly, P.J. (2002). "Rock strength assessment from geophysical logging." *Proc. 8th Int. Symp. on Borehole Geophysics for Minerals, Geotechnical and Groundwater Applications*, Toronto, Ontario, 21st–23rd August.

Huang, S.L., and Wang, Z.W. (1997). "The mechanics of diamond core drilling of rocks." *Int. J. Rock Mech. Min. Sci.*, 34(3–4), 134e1–134e14.

Hurtado, J.E., Londono, J.M., and Meza, M.A. (2001). "On the applicability of neural networks for soil dynamic amplification analysis." *Soil Dyn. Earthquake Eng.*, 21, 579–591.

Jung, S.J., Prisbrey, K., and Wu, G. (1994). "Prediction of rock hardness and drillability using acoustic emission signatures during indentation." *Int. J. Rock Mech. Min. Sci. Geomech. Abstr.*, 31(5), 561–567.

Kahraman S., Korkmazve, S., and Akcay, M. (1996). "The reliability of using Schmidt hammer and point load strength test in assessing uniaxial compressive strength." *K.T.U. Dept. Geol. Eng. 30th year Symp. Book*, Trabzon, 362–369.

Kahraman, S., Balci, C., Yazici, S., and Bilgin, N. (2000). "Prediction of the penetration rate of rotary blasthole drills using a new drillability index." *Int. J. Rock Mech. Min. Sci.*, 37(5), 729–743.

Kahraman, S. (2001). "Evaluation of simple methods for assessing the uniaxial compressive strength of rock." *Int. J. Rock Mech. Min. Sci.*, 38(7), 981–994.

Kahraman, S. (2002). "Estimating the direct P-wave velocity value of intact rock from indirect laboratory measurements." *Int. J. Rock Mech. Min. Sci.*, 39, 101–104.

Kahraman, S., Altun, H., Tezekici, B.S., and Fener, M. (2006). "Sawability prediction of carbonate rocks from shear strength parameters using artificial neural networks." *Int. J. Rock Mech. Min. Sci.*, 43, 157–164.

Kahraman, S., and Alber, M. (2006). "Predicting the physico-mechanical properties of rocks from electrical impedance spectroscopy measurements." *Int. J. Rock Mech. Min. Sci.*, 43(4), 543–553.

Kahraman, S., Gunaydin, O., Alber, M., and Fener, M. (2009). "Evaluating the strength and deformability properties of Misis fault breccia using artificial neural networks." *Expert Syst. Appl.*, 36, 6874–6878.

Karakus, M., Kumral, M., and Kilic, O. (2005). "Predicting elastic properties of intact rocks from index tests using multiple regression modelling." *Int. J. Rock Mech. Min. Sci.*, 42, 323–330.

Katz, O., Reches, Z., and Roegiers, J.C. (2000). "Evaluation of mechanical rock properties using Schmidt Hammer." *Int. J. Rock Mech. Min. Sci.*, 37(4), 723–728.

Kayadelen, C. (2008). "Estimation of effective stress parameter of unsaturated soils by using artificial neural networks." *Int. J. Numer. Anal. Methods Geomech.*, 32(9), 1087–1106.

Kidybinski, A. (1980). "Bursting liability indices of coal." *Int. J. Rock Mech. Min. Sci. Geomech. Abstr.*, 17, 167–171.

Kilic, A., and Teymen, A. (2008). "Determination of mechanical properties of rocks using simple Methods." *Bull. Eng. Geol. Environ.*, 67, 237–244.

Kim, K., and Gao, H., (1995). "Probabilistic approaches to estimating variation in the mechanical properties of Rock masses." *Int. J. Rock Mech. Min. Sci. Geomech. Abstr.*, 32(2), 111–120.

Knill, J.L., Franklin, J.A., and Malone, A.W. (1968). "A study of acoustic emission from stressed rock." *Int. J. Rock Mech. Min. Sci. Geomech. Abstr.*, 16, 23–35.

Lai, S., and Sera, M. (1997). "Concrete strength prediction by mean of neural networks." *Constr. Build. Mater.*, 11(2), 93–98.

Lee, I.M., and Lee, J.H. (1996). "Prediction of pile bearing capacity using artificial neural networks." *Comput. Geotech.*, 18(3), 189–200.

Lee, S.J., Lee, S.R., and Kim, Y.S. (2003). "An approach to estimate unsaturated shear strength using artificial neural network and hyperbolic formulation." *Comput. Geotech.*, 30, 489–503.

McNally, G.H. (1990). "The prediction of geotechnical rock properties from sonic and neutron logs." *Explor. Geophys.*, 21, 65–71.

Miklusova, V., Usalova, L., Ivanicova, L., and Krepelka, F. (2006). "Acoustic signal—new feature in monitoring of rock disintegration process." *Contrib. Geophys. Geodesy*, SAS 36, SI 6SGK, 125–133.

Miller, D.M., Kaminsky, E.J., and Rana, S. (1995). "Neural network classification of remote-sensing data." *Comput. Geosci.*, 21(3), 377–386.

Mogi, K. (1973). "Rock fracture." *Annu. Rev. Earth Planet. Sci.*, 1, 63–84.

Moulenkamp, F., and Grima, M.A. (1999). "Application of neural networks for the prediction of unconfined compressive strength (UCS) for Equotip hardness." *Int. J. Rock Mech. Min. Sci.*, 36, 29–39.

Obert, L. (1941). "Use of sub audible noise for prediction of rock bursts—Part I." *U.S. Bureau of Mines*, R.I. 3555.

Obert, L., and Duvall, W.I. (1942). "Use of sub audible noise for prediction of rock bursts—Part II." *U.S. Bureau of Mines*, R.I. 3654.

O'Rourke, J.E. (1989). "Rock index properties for geo engineering underground development." *Min. Eng.*, 106–110.

Rafiq, M.Y, Bugmann, G., and Easterbrook, D.J. (2001). "Neural network design for engineering applications." *Comput. Struct.*, 79, 1541–1552.

Rasmussen, C.E. (2000). *The Infinite Gaussian Mixture Model. Neural Information Processing Systems.* Cambridge, MA, MIT Press. 12, 554–560.

Romeo, G., Mele, F., and Morelli, A. (1995). "Neural networks and discrimination of seismic signals." *Comput. Geosci.*, 21(2), 279–288.

Sachapazis, C.I. (1990). "Correlating Schmidt hardness with compressive strength and Young's modulus of carbonate rocks." *Bull. Int. Assoc. Eng. Geol.*, 42, 75–83.

Sarkar, K., Tiwary, A., and Singh, T.N. (2010). "Estimation of strength parameters of rock using artificial neural networks." *Bull. Eng. Geol. Environ.*, 69(4), 599–606.

Schon, J.H. (1996). "Physical properties of rock—fundamentals and principles of petrophysics." *Handbook of Geophysical Exploration, Seismic Exploration*, Pergamon Press, Oxford, p. 583.

Shamina, O.G. (1975). "Modeling of earthquakes." *Earth Phys.*, 10, 10–20.

Sheorey, P.R., Barat, D., Das, M.N., Mukherjee, K.P., and Singh, B. (1984). "Schmidt hammer rebound data for estimation of large scale in situ coal strength." *Int. J. Rock Mech. Min. Sci.*, 21, 39–42.

Singh, M., and Rao, S.K. (2005). "Empirical methods to estimate the strength of the jointed rock masses." *Eng. Geol.*, 77(1–2), 127–137.

Singh, R.N., Hassani, F.P., and Elkington, P.A.S. (1983). "The application of strength and deformation index testing to the stability assessment of coal measures excavations." *Proc., 24th US Symp. on Rock Mech.*, Texas A & M University, AEG, 599–609.

Singh, V.K., Singh, D., and Singh, T.N. (2001). "Prediction of strength properties of some schistose rocks from petrographic properties using artificial neural networks." *Int. J. Rock Mech. Min. Sci.*, 38, 269–284.

Sinha, S.K., and Wang, M.C. (2007). "Artificial neural network prediction models for soil compaction and permeability." *Geotech. Geol. Eng.*, 26, 47–64.

Sonmez, H., Tuncay, E., and Gokceoglu, C. (2004). "Models to predict the uniaxial compressive strength and the modulus of elasticity for Ankara Agglomerate." *Int. J. Rock Mech. Min. Sci.*, 41, 717–729.

Sonmez, H., Gokceoglu, C., Nefeslioglu, H.A., and Kayabasi, A. (2006). "Estimation of rock modulus—for intact rocks with an artificial neural network and for rock masses with a new empirical equation." *Int. J. Rock Mech. Min. Sci.*, 43, 224–235.

Taskiran, T. (2010). "Prediction of California bearing ratio (CBR) of fine grained soils by AI methods." *Adv. Eng. Softw.*, 41, 886–892.

Tiryaki, B. (2008a). "Application of artificial neural networks for predicting the cuttability of rocks by drag tools." *Tunnelling Underground Space Technol.*, 23, 273–280.

Tiryaki, B. (2008b). "Predicting intact rock strength for mechanical excavation using multivariate statistics, artificial neural networks and regression trees." *Eng. Geol.*, 99, 51–60.

Tugrul, A., and Zarif, I.H. (1999). "Correlation of mineralogical and textural characteristics with engineering properties of selected granitic rocks from Turkey." *Eng. Geol.*, 51, 303–317.

Vardhan, H., and Murthy, Ch.S.N. (2007). "An experimental investigation of jack hammer drill noise with special emphasis on drilling in rock s of different compressive strengths." *Noise Control Eng. J.*, 55(3), 282–293.

Vardhan, H., Adhikari, G.R., and Govindaraj, M. (2009). "Estimating rock properties using sound levels produced during drilling." *Int. J. Rock Mech. Min. Sci.*, 46(3), 604–612.

Vukelic, M.A., and Miranda, E.N. (1996). "Neural networks in petroleum engineering." *Int. J. Neural Syst.*, 7, 187.

Ward, B. (1998). German Creek Mines Rock Strength from Velocity Logs. Unpublished report for Capricorn Coal Management Pty Ltd.

Williams, E. and Hogan, P.C. (2006). "Monitoring acoustic emission levels with changes in rock cutting conditions." *Pub., University of New South Wales*, www.mining.unsw.edu.au/Publications/publications_staff/Paper_WilliamsHogan_AE_2006_web.htm, (July 2006).

Yagiz, S. (2008). "Utilizing rock mass properties for predicting TBM performance in hard rock condition." *Tunn Undergr Sp Technol.*, 23(3), 326–339.

Yagiz, S., Sezer, E.A., and Gokceoglu, C. (2011). "Artificial neural networks and non-linear regression techniques to assess the influence of slake durability cycles on the prediction of uniaxial compressive strength and modulus of elasticity for carbonate rocks." *Int. J. Numer. Anal. Methods Geomech.*, DOI: 10.1002/nag.1066.

Yang, Y., and Zhang, Q. (1998). "The applications of neural networks to rock engineering systems." *Int. J. Rock Mech. Min. Sci.*, 35(6), 727–745.

Yasar, E., and Erdogan, Y. (2004a). "Estimation of rock physicomechanical properties using hardness methods." *Eng. Geol.*, 71, 281–288.

Yasar, E., and Erdogan, Y. (2004b). "Correlating sound velocity with the density, compressive strength and Young's modulus of carbonate rocks." *Int. J. Rock Mech. Min. Sci.*, 41(5), 871–875.

Yavuz, H., Ugur, I., and Demirdag, S. (2008). "Abrasion resistance of carbonate rocks use in dimension stone industry and correlation between abrasion and rock properties." *Int. J. Rock Mech. Min. Sci.*, 45(2), 260–267.

Yilmaz, I., and Sendir, H. (2002). "Correlation of Schmidt hardness with unconfined compressive strength and Young's modulus in gypsum from Sivas (Turkey)." *Eng. Geol.*, 66, 211–219.

Yuanyou, X., Yanming, X., and Ruigeng, Z. (1997). "An engineering geology evaluation method based on an artificial neural network and its application." *Eng. Geol.*, 47, 149–156.

Zborovjan, M. (2001). "Identification of minerals during drilling process via acoustic signal." *Metall. Foundry*, 26(4), Krakow, Poland.

Zborovjan, M., Lesso, I., and Dorcak, L. (2003). "Acoustic identification of rocks during drilling process." *J. Acta Montanistica Slovaca*, 8(4), 191–193.

Zorlu, K., Gokceoglu, C., Ocakoglu, F., Nefeslioglu, H.A., and Acikalin, S. (2008). "Prediction of uniaxial compressive strength of sandstones using petrography based models." *Eng. Geol.*, 96, 141–158.

5

Equipment for Drilling, Measurement of Sound, and Physico-mechanical Properties of Rocks

5.1 Drilling Machine

In this work, rock drilling operations were performed on BMV 45 T20, computer numerical controlled (CNC) vertical machining center (Figure 5.1). The experimental set-up was in a fiber-and-glass-paned room of 5 m width, 6 m length, and 9 m height. The important specifications of the CNC machine used were:

Table size 450 mm × 900 mm

Recommended optimum air pressure—6 bar

Power supply—415 V, 3 phase, 50 Hz

Carbide drill bits of shank length 40 mm and diameters of 6, 10, 16, and 20 mm were used for drilling operation. Machine was set to drill 30 mm drill-hole length. Since the drilling method affects the sound produced, an attempt was made to standardize the testing procedure. Throughout the drilling process a relatively constant rotation speed (RPM) and penetration rate (mm/min) were provided in order to obtain consistent data.

5.2 Sound Level Measuring Instrument

The instrument used for sound measurement in this work was Spark 706 from Larson Davis, Inc., Depew, New York (Figure 5.2). The instrument was equipped with a detachable 10.6-mm microphone and 7.6-cm cylindrical mast type preamplifier. The microphone and preamplifier assembly were connected by an integrated 1.0-m cable. A Larson Davis CAL 200 Precision

FIGURE 5.1
BMV 45 T20, CNC vertical machining center.

FIGURE 5.2
Spark 706, sound measuring instrument.

Acoustic Calibrator was used for calibrating the sound level meter. Before taking any measurement, the acoustical sensitivity of the sound level meter was checked using the calibrator.

5.3 Compression Testing Machine

Compressive strength is one of the most important mechanical properties of rock material, used in blast-hole design. In this work, AIM-317E-Mu micro controller compression testing machine was used for measurement of uniaxial compressive strength (Figure 5.3). It had an intelligent pace rate controller, motorized pumping unit, and loading unit with maximum loading capacity of 2000 kN.

FIGURE 5.3
AIM-317E-Mu, compression testing machine.

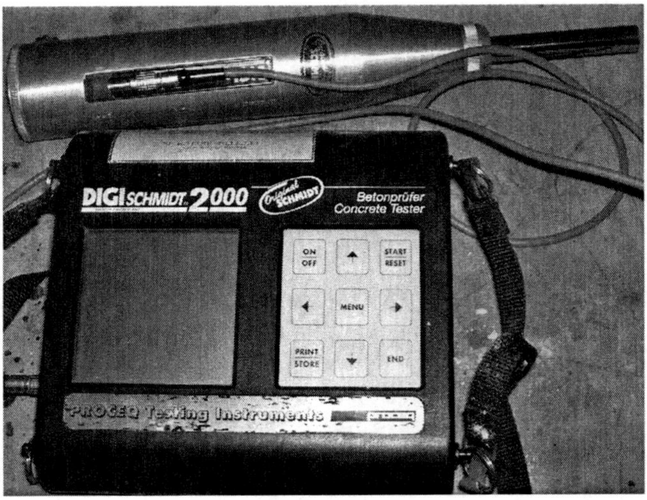

FIGURE 5.4
DIGI-SCHMIDT 2000, SRN measuring instrument.

5.4 Schmidt Hammer

Schmidt hammer rebound hardness is often measured during the early part of field investigation. It is a measure of the hardness of the rock material by counting the rebound degree. In this work, tests were performed with DIGI–SCHMIDT 2000 (Figure 5.4). The graphic LCD 128 × 128 pixel display unit of the instrument immediately displays the rebound value.

5.5 P-Wave Velocity Measuring Equipment

The velocity of ultrasonic pulses traveling in a solid material depends on the density and elastic properties of that material. The quality of some materials is sometimes related to their elastic stiffness so that measurement of ultrasonic pulse velocity in such materials can often be used to indicate their quality, as well as to determine their elastic properties. In this work, the Portable Ultrasonic Non-Destructive Digital Indicating Tester (PUNDIT–6 PC1000) with pulse generator operated at 500 V and pulse repetition frequency of 10 pulses per second was used to measure the propagation speed of a pulse of ultrasonic longitudinal stress waves (Figure 5.5).

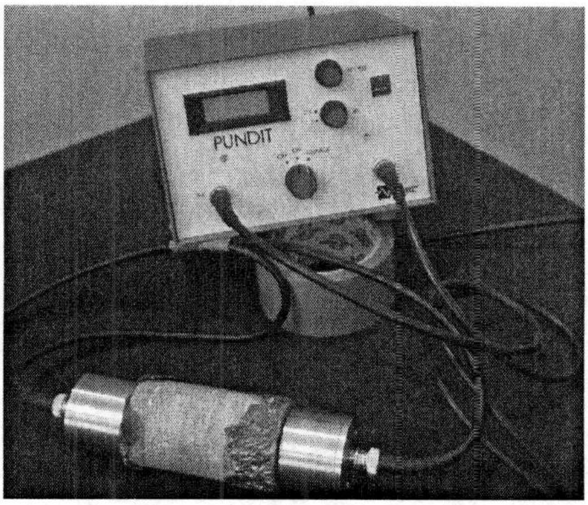

FIGURE 5.5
PUNDIT-6 PC1000, *P*-wave velocity measuring instrument.

FIGURE 5.6
Brazilian tensile testing machine.

5.6 Tensile Strength Testing Machine

Rock material generally has a low tensile strength. The low tensile strength is due to the existence of micro cracks in the rock. The existence of micro cracks may also be the cause of rock failing suddenly in tension with a small strain. In this work, tensile strength of rock was obtained from Brazilian test loading frame with 100 kN capacity, having a base and a cross head joined together with two solid pillars with nuts (Figure 5.6). At the top, the pillars have long threads for height adjustment and on the base, a 100-kN hydraulic jack is centrally fixed between the pillars. This jack has an integral pumping unit and oil reservoir. A 100-kN capacity pressure gauge is fixed to the jack for indicating the load on the specimen and an operating handle is provided with the jack.

6

Measurement of Rock Properties and Sound Level

The aim of the work presented in this book was to develop the prediction models to predict various rock properties from sound levels produced during rotary drilling. To achieve this goal, 14 types of igneous rocks were collected from different localities in India, making sure there was a variety of strength represented. During sample collection, each block was inspected for macroscopic defects so that it would provide test specimens free from fractures and joints.

6.1 Rock Compressive Strength

To determine the uniaxial compressive strength (UCS) of the rock samples, 54-mm-diameter NX-size core specimens, having a length-to-diameter ratio of 2.5:1, were prepared as per ISRM suggestions (Ulusay and Hudson 2007). Each block was represented by at least five core specimens. The oven-dried and NX-size core specimens were tested by using a microcontroller compression testing machine. The basic statistical evaluations of UCS such as mean, maximum, minimum, and standard deviations of different igneous rocks are given in Table 6.1. However, only mean values of UCS were considered for developing the models.

6.2 Schmidt Rebound Number

In this work, Schmidt rebound number (SRN) tests were performed with the Schmidt hammer. All tests were performed with the hammer held vertically downward and at right angles to the horizontal rock face (Aydin 2009). To get Schmidt hammer rebound number, 10 readings were taken on a particular rock sample and then the mean of nearer values were used for the analysis. The basic statistical evaluations of SRN of different igneous rocks are given in Table 6.1.

TABLE 6.1

Statistical Values of Igneous Rock Properties and Range of A–Weighted Equivalent Sound Level Values Obtained During Drilling

Rock Type		Koira Grey Granite	Quartz Monzonite	Granodiorite	Peridotite	Serpentine	Syenite	Norite	Granite Porphyry	Pegmatite	Charnockite	Diorite Porphyry	Grey Granite	Dolerite	Gabbro
UCS (MPa)	Min	77.6	41.6	79.0	64.3	37.0	46.1	48.3	51.0	34.9	66.2	56.7	46.0	65.8	101.8
	Max	77.9	42.5	80.1	65.2	37.0	48.1	49.0	51.7	36.1	67.3	57.9	46.9	66.8	102.6
	SD	0.123	0.339	0.505	0.367	0.000	0.718	0.255	0.292	0.524	0.474	0.447	0.354	0.412	0.292
	Mean	77.8	42.0	79.6	64.9	37.0	47.0	48.7	51.2	35.4	66.8	57.2	46.4	66.1	102.2
SRN	Min	55	37	57	50	34	40	42	42	33	50	47	40	51	64
	Max	58	41	59	54	37	45	44	46	36	55	50	42	55	72
	SD	1.225	1.581	0.707	1.871	1.225	2.739	0.707	1.414	1.225	1.871	1.225	0.707	1.414	2.916
	Mean	57	39	58	52	36	42	43	44	35	53	48	41	53	68
Density (gm/cc)	Min	2.697	2.505	2.399	2.613	2.457	2.494	2.529	2.716	2.482	2.623	2.590	2.557	2.601	3.101
	Max	2.852	2.711	2.576	2.675	2.564	2.568	2.591	2.844	2.521	2.885	2.641	2.580	2.744	3.199
	SD	0.071	0.088	0.065	0.025	0.041	0.030	0.022	0.047	0.015	0.108	0.019	0.009	0.051	0.038
	Mean	2.773	2.556	2.481	2.649	2.512	2.536	2.558	2.766	2.496	2.699	2.615	2.571	2.665	3.168
P-wave velocity (km/sec)	Min	6.083	5.498	6.298	5.892	5.498	5.594	5.432	5.638	5.412	6.002	5.785	5.532	5.912	6.809
	Max	6.165	5.596	6.344	5.912	5.558	5.698	5.987	5.757	5.574	6.093	5.823	5.643	6.159	6.951
	SD	0.029	0.044	0.017	0.007	0.022	0.037	0.204	0.044	0.058	0.034	0.014	0.04	0.099	0.05
	Mean	6.127	5.543	6.323	5.901	5.527	5.645	5.651	5.685	5.485	6.037	5.805	5.591	5.993	6.879

Tensile strength (MPa)	Min	9.58	4.69	9.79	7.75	4.02	5.16	5.39	6.05	3.79	7.95	6.78	5.18	7.89	12.58
	Max	9.61	4.79	9.93	7.91	4.28	5.32	5.65	6.34	4.35	8.15	6.95	5.23	7.98	12.64
	SD	0.012	0.041	0.058	0.058	0.101	0.069	0.112	0.110	0.223	0.100	0.074	0.021	0.039	0.024
	Mean	9.60	4.72	9.83	7.82	4.16	5.28	5.47	6.17	3.98	8.05	6.89	5.21	7.96	12.62
Young's modulus (GPa)	Min	73.6	65.4	73.9	71.4	64.1	68.3	69	69.5	63.2	71.7	70.1	68.3	71.6	75
	Max	74.0	66.1	74.28	71.9	64.5	68.9	69.9	70.0	63.8	72.3	70.8	68.8	72.2	75.5
	SD	0.141	0.292	0.152	0.187	0.200	0.255	0.324	0.187	0.245	0.224	0.332	0.187	0.224	0.235
	Mean	73.8	65.7	74.1	71.7	64.3	68.7	69.4	69.8	63.4	72.0	70.5	68.5	71.9	75.2
Porosity (%)	Min	0.4789	2.3169	0.4101	0.8991	2.5987	1.6098	1.4390	1.3588	2.8120	0.8423	1.1860	1.5643	0.8889	0.1896
	Max	0.5001	2.3400	0.4290	0.9534	2.6900	1.6402	1.4730	1.3882	2.9355	0.9267	1.2260	1.7070	0.8999	0.1994
	SD	0.0090	0.0095	0.0075	0.0208	0.0348	0.0112	0.0136	0.0113	0.0450	0.0304	0.0146	0.0601	0.0045	0.0045
	Mean	0.4854	2.3316	0.4211	0.9349	2.6528	1.6284	1.4578	1.3736	2.8755	0.8792	1.2058	1.6665	0.8922	0.1956
L_{eq} (dB)	Min	112.4	93.9	113.1	107.7	90.7	96.6	97.5	99.1	90.0	109.0	104.0	95.8	107.9	118.2
	Max	118.1	97.6	118.9	112.4	93.8	100.1	100.9	102.9	94.2	114.1	108.3	99.6	112.8	121.3

6.3 Dry Density of Rock (ρ)

Density is a measure of mass per unit of volume. Density of rock material varies and is often related to the porosity of the rock. It is sometimes defined by unit weight and specific gravity. In this work, the density of each core sample was measured after the removal of moisture from it. The moisture was removed by placing the samples in an electric oven at approximately 80° C for 1 hour, and they were dried at room conditions (Ulusay and Hudson 2007). The density data of the dry sample were obtained from the measurement of volume and mass of each sample and using the following equation:

$$\rho(g/cc) = \frac{Mass\ of\ sample}{Volume\ of\ sample}$$

Each test was repeated five times and the average values were recorded. The average results of dry densities of different rocks were considered for analysis. The basic statistical evaluations of dry densities of different igneous rocks are given in Table 6.1.

6.4 P-Wave Velocity

To determine the P-wave velocity of different rocks, rock blocks were cored in the laboratory for NX size core recovery (Ulusay and Hudson 2007). A pulse of ultrasonic (>20 kHz) longitudinal stress wave was introduced into the rock core from one end by a transducer coupled to the surface with a coupling gel or grease. The pulse travels through the rock core and was received by a similar transducer coupled to the opposite surface. The transit time of the pulse is determined by the instrument. The distance between the transducers is divided by the transit time to obtain the pulse velocity. Each test was repeated five times. The basic statistical evaluations of P-wave velocities of different igneous rocks are given in Table 6.1. Only mean values of P-wave velocities were considered for developing the models.

6.5 Rock Tensile Strength

To determine the Brazilian tensile strength of the rock samples, 54 mm diameter NX-size core specimens, having a thickness less than 27 mm were prepared as per ISRM suggestions (Ulusay and Hudson 2007). The cylindrical surfaces were made free from any irregularities across the thickness using polishing machine. End faces were made flat to within 0.25 mm and parallel to within 0.25°. The specimen was wrapped around its periphery with one layer of the masking tape and loaded into the Brazil tensile test apparatus across its diameter. Load was applied continuously at a constant rate such that failure occurs within 15–30 seconds. Ten specimens of the same sample were tested. The basic statistical evaluations of Brazilian tensile strength of different igneous rocks are given in Table 6.1. Only mean values of Brazilian tensile strength were considered for developing the models.

6.6 Young's Modulus (E) and Percentage Porosity of the Rock

Young's modulus is modulus of elasticity measuring the stiffness of a rock material. For small strains, it is defined as the ratio of the rate of change of stress with strain. Porosity describes how densely the material is packed. Young's modulus and porosity values, determined by the rock supplier, were used during analysis as the instrumentation to determine these properties were not available during the work. The basic statistical evaluations of Young's modulus and percentage porosity of different igneous rocks are given in Table 6.1. Only mean values of Young's modulus and percentage porosity were considered for developing the models.

6.7 A-Weighted Equivalent Sound Level

For rotary drilling, rock blocks with dimensions of 20 cm × 20 cm × 20 cm were prepared by sawing from block samples. During drilling, to overcome the vibration, the rock blocks were firmly held by a vice, which was kept on the table of the machine. Sound level measurements were carried out for rotation speeds of 150, 200, 250, and 300 RPM, and penetration rates of 2, 3, 4, and 5 mm/min on each rock block. Since CNC machine was used in this work, a numerical controlled (NC) program was written for drilling operation. In the program itself drill speed and penetration rate were fed and controlled.

For each combination of drill bit diameter, drill bit speed, and penetration rate, a total of 64 sets of test conditions were arrived (drill bit diameter of 6, 10, 16, and 20 mm; drill bit speed of 150, 200, 250, and 300 RPM; penetration rate of 2, 3, 4, and 5 mm/min). An A-weighted equivalent continuous sound level (L_{eq}) was recorded for all 64 different drill holes of 30 mm depth on each rock block.

For all measurements, the sound level meter was kept at a distance of 1.5 cm from the periphery of the drill bit (Figure 6.1). For a particular condition and for the same rock block, the sound level was measured three times in relatively rapid succession. It was found that the recorded equivalent sound levels were almost consistent. The arithmetic average of each set of three measurements was computed to yield an average A-weighted equivalent sound level for a particular condition. Range (maximum and minimum) of A-weighted equivalent sound level recorded during drilling of igneous rocks are given in Table 6.1.

For the igneous rocks, the variation of equivalent sound level (L_{eq}) produced during drilling with various rock properties at drill bit diameter of 20 mm, penetration rate of 5 mm/min, and drill bit speed of 300 rpm are shown in Figure 6.2 to Figure 6.8.

From the above figures, it can be observed that with increase in properties like UCS, SRN, density, P-wave velocity, tensile strength, and Young's modulus values of the igneous rocks, the A-weighted equivalent sound level produced during rock drilling increases, where as for porosity values, it decreases.

For 15 minutes, the sound level was measured at 1.5 cm from the drill bit without drilling. The equivalent sound level of 65.2 dB was recorded without drilling, which was mainly due to the noise of the CNC machine.

It may be argued that sound produced from the CNC machine itself may affect the sound level measurement during rock drilling. It is important to mention here that if the sound level difference between two sources is more than 10 dB, then the total sound level will remain the same as that of the higher source (Colin, 2005). Further, taking the measurement very close to the source will reduce the effect of sound produced from other sources.

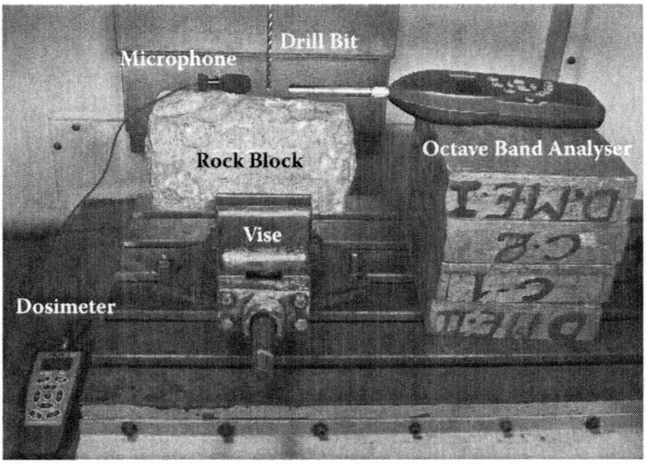

FIGURE 6.1
Position of microphone from drill setup.

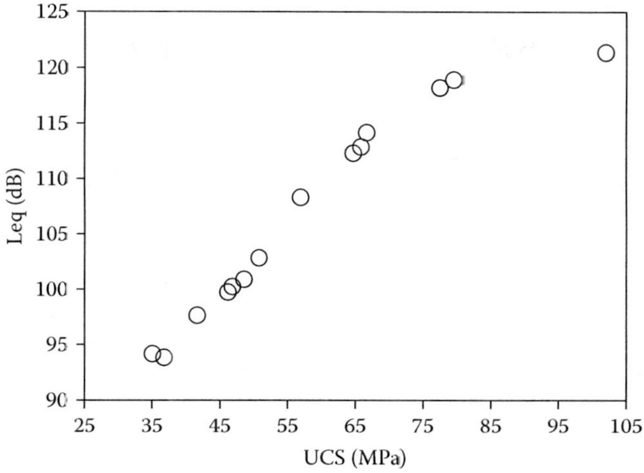

FIGURE 6.2
Variation of A-weighted equivalent sound level produced during rock drilling vis-à-vis UCS of the drilled igneous rock.

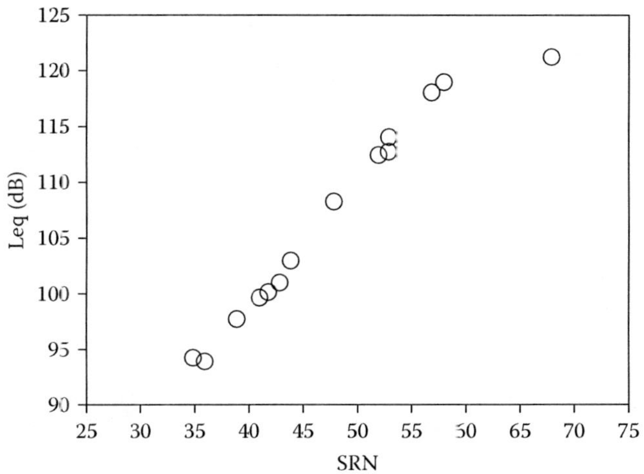

FIGURE 6.3
Variation of A-weighted equivalent sound level produced during rock drilling vis-à-vis SRN of the drilled igneous rock.

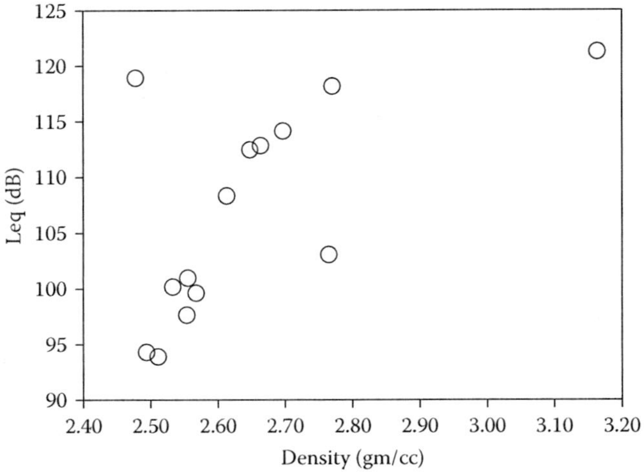

FIGURE 6.4
Variation of A-weighted equivalent sound level produced during rock drilling vis-à-vis Dry density of the drilled igneous rock.

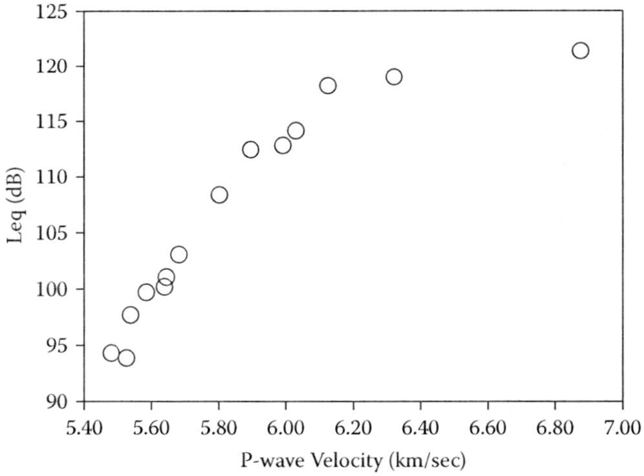

FIGURE 6.5
Variation of A-weighted equivalent sound level produced during rock drilling vis-à-vis P-wave velocity of the drilled igneous rock.

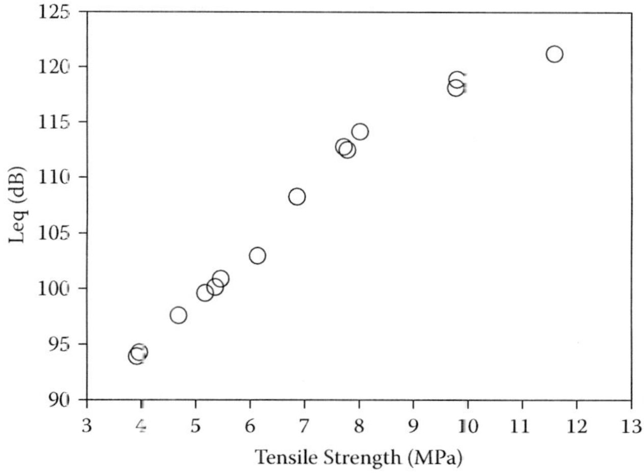

FIGURE 6.6
Variation of A-weighted equivalent sound level produced during rock drilling vis-à-vis tensile strength of the drilled igneous rock.

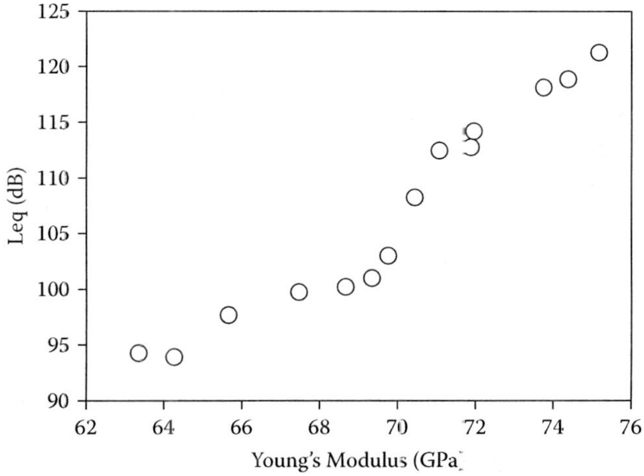

FIGURE 6.7
Variation of A-weighted equivalent sound level produced during rock drilling vis-à-vis Young's modulus of the drilled igneous rock.

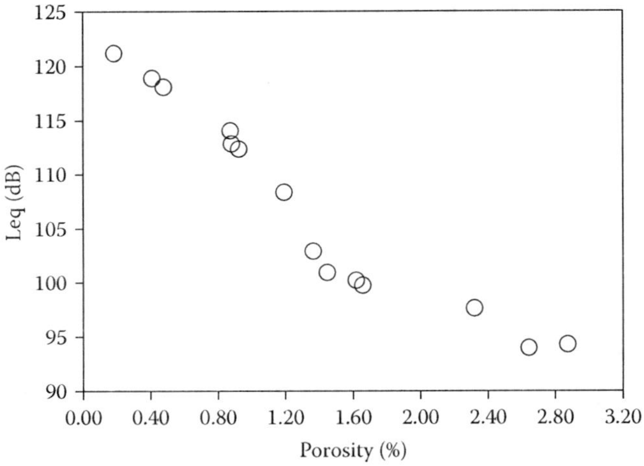

FIGURE 6.8
Variation of A-weighted equivalent sound level produced during rock drilling vis-à-vis percentage porosity of the drilled igneous rock.

References

Aydin, A. (2009). "ISRM suggested method for determination of Schmidt hammer rebound hardness: Revised version." *Int. J. Rock Mech. Min. Sci.*, 46, 627–634.

Colin, H. (2005). *Noise Control from Concept to Application*. New York, Taylor & Francis.

Ulusay, R., and Hudson, J.A., eds., (2007). "The complete ISRM suggested methods for rock characterization, testing and monitoring: 1974–2006." *Compilation Arranged by the ISRM Turkish National Group*, Ankara, Turkey, 628 p.

7

Regression Modeling

7.1 Introduction

To obtain applicable and practical predictive qualitative relationships, it is necessary to model the rock properties and the drill process variables. Multiple regression analysis is widely used for modeling and analyzing the experimental results. Regression is the determination of the statistical relationship between two or more variables. It is the statistical method to deal with the formulation of mathematical model depicting the relationship among the variables which can be used for the purpose of prediction of values of dependent or response variables, given the values of predictor or independent variable(s), the analysis concerning the relationship is known as multiple correlation and equations describing such relationships are called as multiple regression equations. Physical properties of rocks can be predicted by multiple regression modeling, the statistical methodology used to relate variables (Bowerman and O'Connell 1990).

In order to establish the predictive models among the parameters obtained during experimental work presented in this book, multiple regression and analysis of variance (ANOVA) techniques were used. For modeling and analysis Minitab 15 software for windows was used.

7.2 Multiple Regression Analysis and ANOVA Technique

The laboratory experimental results obtained in this work were used to model the various responses using multiple regression method by using a nonlinear fit among the responses and the corresponding significant parameters. The performance of the model depends on a large number of factors that act and interact in a complex manner. When the predictor variables in a multiple regression model are interrelated or are dependent on each other, a multicollinearity problem exists and hinders the ability to assess the importance of a predictor variable. The solutions to the problem are to remove one or more of the highly correlated predictor variables or to add more scenarios used in building the model.

The mathematical modeling of sound level produced during drilling is influenced by many factors. Therefore, a detailed process representation anticipates a second-order model. ANOVA was carried out to find which input parameter significantly affects the desired response. To facilitate experiments and measurement, four important predictor variables were considered in this work. They are drill bit diameter in mm (A), drill bit speed in RPM (B), penetration rate in mm/min (C), and equivalent sound level produced during drilling in dB (D). The responses considered were uniaxial compressive strength (UCS), Schmidt rebound number (SRN), dry density (ρ), P-wave velocity (V_P), tensile strength (TS), Young's modulus (E), and percentage porosity (n). The mathematical models for the physico-mechanical properties with parameters under consideration can be represented by

$$Y = f(x_1, x_2, x_3, \ldots) + \in \tag{7.1}$$

where Y is the response, x_1, x_2, x_3 are the independent process variables, and $\in$ is fitting error. A quadratic model of f can be written as

$$f = b_0 + \sum_{i=1}^{n} b_i x_i + \sum_{i=1}^{n} b_{ij} x_i^2 + \sum_{i<j}^{n} b_{ij} x_i x_j + \in \tag{7.2}$$

where b_i represents the linear effect of x_i, b_{ij} represents the quadratic effect of x_i, and b_{ij} in fourth term represents linear interaction between x_i and x_j. Then, the regression models contain linear terms, squared terms, and cross-product terms.

Individual responses were modeled in this work by using multiple regression analysis. The coefficient table lists the estimated coefficients for all the predictors. p-value determines the observed relationship between response and the predictors and indicates whether it is statistically significant. If the p-value is less than the selected α level, the association is statistically significant and the model was selected. (To test the significance, one needs to set a risk level called the *alpha level*. In most cases, the "rule of thumb" is to set the alpha level at 0.05, i.e., the 95% confidence interval.) On the other hand, if it is more than the selected α level, it is not statistically significant, and the independent variable under consideration is removed from the model, and the regression analysis is performed by using all the remaining independent variables. The procedure was continued by removing one independent variable at a time from the model. The screening was stopped when the independent variables remaining in the model could not be removed from the system.

The criterion for the explanation capability of the dependent variable from information obtained via the independent variables is the strength of the relation (R^2). When the explained variation is divided by the total variation of y values, that is, when R^2 is calculated, then the proportion of variation of the dependent variable changes is obtained and that is explained by the independent variable. As the number of independent variables increases, R^2 will be greater. Therefore, R^2 should be adjusted (corrected) as follows (Levin and Rubin 2006):

$$\bar{R}^2 = \left[R^2 - \frac{k-1}{n-1}\right]\left[\frac{n-1}{n-k}\right] \tag{7.3}$$

where, k is the number of model parameters, and n is the number of scenarios. The higher the value of R^2 means that there is a good correlation between the experimentally measured values and the predicted values using the developed models.

7.3 Selection of Rock Samples for Modeling

For experimental works, different categories of rocks were collected. Out of the total collected rocks, there were 14 igneous rock types. Out of these, 10 rock types were used for developing the model and 4 types were used to test the accuracy of the developed model, so in total 640 (i.e., 64 × 10) Equivalent Sound Level (L_{eq}) values were used for developing the models and 256 (i.e., 64u 4) L_{eq} values were used for testing the developed models.

7.4 Regression Models

Multiple regression model to predict uniaxial compressive strength of the igneous rock types is:

$$\begin{aligned} UCS = {} & 468.204 + 2.225 \times A + 0.069 \times B + 2.502 \times C - 10.273 \times D \\ & + 0.061 \times D^2 - 0.025 \times A \times D - 0.001 \times B \times D - 0.028 \times C \times D \end{aligned}$$

Significance of regression coefficients for estimation of uniaxial compressive strength is listed in Table 7.1A, which also shows t value and p value. The final ANOVA table of the reduced quadratic model for UCS is shown in Table 7.1B. This table also represents degrees of freedom (DF), mean square

TABLE 7.1A

Significance of Regression Coefficients for the Estimation of Uniaxial Compressive Strength (Igneous Rocks)

Model Terms for UCS	Parameter Estimate (Coefficients)	*t*-Value	*p*-Value
Constant	468.204	22.692	0.000
A	2.225	8.048	0.000
B	0.069	2.671	0.008
C	2.502	1.948	0.042
D	−10.273	−26.767	0.000
D^2	0.061	33.119	0.000
AD	−0.025	−9.465	0.000
BD	−0.001	−3.149	0.002
CD	−0.028	−2.294	0.022

TABLE 7.1B

Analysis of Variance (ANOVA) for the Selected Quadratic Model for the Estimation of UCS (Igneous Rocks)

Source of Variations	Degree of Freedom	Sum of Squares	Mean Squares	*F*-Value	*p*-Value
Model	8	221064.60	27633.07	3055.87	0.000
Linear	4	6850.91	1712.73	189.41	0.000
Square	1	9918.64	9918.64	1096.88	0.000
Interaction	3	936.54	312.18	34.52	0.000
Residual error	631	5705.90	9.04	—	—
Total	639	226770	—	—	—

TABLE 7.1C

Model Summary for Dependent Variable (UCS—Igneous Rocks)

R^2	Predicted R^2	Adjusted R^2	Standard Error
0.9748	0.9739	0.9745	3.00710

(MS), sum of squares (SS), *F*-value, and *p*-value associated with factors. As seen from Table 7.1C, for igneous rocks, the selected model explains 97.45% of the total variation in the observed UCS tests.

Multiple regression model to predict Schmidt rebound number of the igneous rock types is:

$$SRN = 123.203 + 0.592 \times A + 0.020 \times B - 2.565 \times D + 0.018 \times D^2 - 0.008 \times A \times D$$

Significance of regression coefficients for estimation of Schmidt rebound number is listed in Table 7.2A, which also shows *t*-value and *p*-value. The final

TABLE 7.2A

Significance of Regression Coefficients for the Estimation of Schmidt Rebound Number (Igneous Rocks)

Model Terms for SRN	Parameter Estimate (Coefficients)	*t*-Value	*p*-Value
Constant	123.203	14.013	0.000
A	0.592	4.964	0.000
B	0.020	1.776	0.046
D	−2.565	−15.504	0.000
D^2	0.018	22.432	0.000
AD	−0.008	−6.651	0.000

TABLE 7.2B

Analysis of Variance (ANOVA) for the Selected Quadratic Model for the Estimation of Schmidt Rebound Number (Igneous Rocks)

Source of Variations	Degree of Freedom	Sum of Squares	Mean Squares	*F*-Value	*p*-Value
Model	5	55380.84	9230.140	5474.98	0.000
Linear	3	426.46	142.154	84.32	0.000
Square	1	848.30	848.300	503.18	0.000
Interaction	1	83.23	41.615	24.68	0.000
Residual error	633	1067.16	1.686	—	—
Total	639	56448.0	—	—	—

TABLE 7.2C

Model Summary for Dependent Variable (SRN—Igneous Rocks)

R^2	Predicted R^2	Adjusted R^2	Standard Error
0.9811	0.9805	0.9809	1.29841

ANOVA table of the reduced quadratic model for SRN is shown in Table 7.2B. This table also represents degrees of freedom (DF), mean square (MS), sum of squares (SS), *F*-value, and *p*-value associated with factors. As seen from Table 7.2C, for igneous rocks, the selected model explains 98.09% of the total variation in the observed SRN tests.

Multiple regression model to predict dry density of the igneous rock types is:

$$\rho = 11.0892 + 0.0387 \times A - 0.1813 \times D + 0.0010 \times D^2 - 0.0004 \times A \times D$$

Significance of regression coefficients for estimation of dry density is listed in Table 7.3A, which also shows *t*-value and *p*-value. The final ANOVA table of the reduced quadratic model for dry density is shown in Table 7.3B.

TABLE 7.3A

Significance of Regression Coefficients for the Estimation of Dry Density (Igneous Rock)

Model Terms for Dry Density	Parameter Estimate (Coefficients)	t-Value	p-Value
Constant	11.0892	19.120	0.000
A	0.0387	4.814	0.000
D	−0.1813	−16.284	0.000
D^2	0.0010	17.966	0.000
AD	−0.0004	−5.205	0.000

TABLE 7.3B

Analysis of Variance (ANOVA) for the Selected Quadratic Model for the Estimation of Dry Density (Igneous Rock)

Source of Variations	Degree of Freedom	Sum of Squares	Mean Squares	F-Value	p-Value
Model	4	16.93505	4.23376	552.05	0.000
Linear	2	2.06246	1.03123	134.47	0.000
Square	1	2.47550	2.47550	322.79	0.000
Interaction	1	0.20780	0.20780	27.10	0.000
Residual error	635	4.86991	0.00767	—	—
Total	639	21.8050	—	—	—

TABLE 7.3C

Model Summary for Dependent Variable (Dry Density—Igneous Rock)

R^2	Predicted R^2	Adjusted R^2	Standard Error
0.7767	0.7729	0.7753	0.0875737

This table also represents degrees of freedom (DF), mean square (MS), sum of squares (SS), F-value, and p-value associated with factors. As seen from Table 7.3C, for igneous rocks, the selected model explains 77.53% of the total variation in the observed dry density tests.

Multiple regression model to predict P-wave velocity of the igneous rock types is:

$$V_P = 36.2677 + 0.1037 \times A + 0.0029 \times B + 0.1094 \times C - 0.5632 \times D + 0.0026 \times D^2 - 0.0009 \times A \times D - 0.0010 \times C \times D$$

Significance of regression coefficients for estimation of P-wave velocity is listed in Table 7.4A, which also shows t-value and p-value. The final ANOVA table of the reduced quadratic model for P-wave velocity is shown in

TABLE 7.4A

Significance of Regression Coefficients for the Estimation of P-Wave Velocity—Igneous Rock

Model Terms for P-Wave Velocity	Parameter Estimate (Coefficients)	t-Value	p-Value
Constant	36.2677	44.786	0.000
A	0.1037	9.563	0.000
B	0.0029	2.886	0.004
C	0.1094	2.169	0.030
D	−0.5632	−37.390	0.000
D^2	0.0026	35.480	0.000
AD	−0.0009	−8.825	0.000
CD	−0.0010	−1.996	0.046

TABLE 7.4B

Analysis of Variance (ANOVA) for the Selected Quadratic Model for the Estimation of P-Wave Velocity—Igneous Rock

Source of Variations	Degree of Freedom	Sum of Squares	Mean Squares	F-Value	p-Value
Model	7	90.3274	11.2909	810.61	0.000
Linear	4	20.2177	5.0544	362.87	0.000
Square	1	17.5345	17.5345	1258.85	0.000
Interaction	2	1.2253	0.4084	29.32	0.000
Residual error	631	8.7892	0.0139	—	—
Total	639	99.1166	—	—	—

TABLE 7.4C

Model Summary for Dependent Variable (P-Wave Velocity—Igneous Rock)

R^2	Predicted R^2	Adjusted R^2	Standard Error
0.9113	0.9081	0.9102	0.118021

Table 7.4B. This table also represents degrees of freedom (DF), mean square (MS), sum of squares (SS), F-value, and p-value associated with factors. As seen from Table 7.4C, for igneous rocks, the selected model explains 91.02% of the total variation in the observed P-wave velocity tests.

Multiple regression model to predict tensile strength of the igneous rock types is:

$$TS = 56.4706 + 0.2730 \times A + 0.0086 \times B + 0.3106 \times C - 1.2657 \times D + 0.0076 \times D^2 - 0.0031 \times A \times D - 0.0001 \times B \times D - 0.0035 \times C \times D$$

Significance of regression coefficients for estimation of tensile strength is listed in Table 7.5A, which also shows t-value and p-value. The final ANOVA table of the reduced quadratic model for tensile strength is shown in Table 7.5B. This table also represents degrees of freedom (DF), mean square (MS), sum of squares (SS), F-value, and p-value associated with factors. As seen from Table 7.5C, for igneous rocks, the selected model explains 97.88% of the total variation in the observed tensile strength tests.

TABLE 7.5A

Significance of Regression Coefficients for the Estimation of Tensile Strength (Igneous Rock)

Model Terms for Tensile Strength	Parameter Estimate (Coefficients)	t-Value	p-Value
Constant	56.4706	22.778	0.000
A	0.2730	8.219	0.000
B	0.0086	2.768	0.006
C	0.3106	2.012	0.045
D	−1.2657	−27.448	0.000
D^2	0.0076	34.392	0.000
AD	−0.0031	−9.783	0.000
BD	−0.0001	−3.295	0.001
CD	−0.0035	−2.394	0.017

TABLE 7.5B

Analysis of Variance (ANOVA) for the Selected Quadratic Model for the Estimation of Tensile Strength (Igneous Rock)

Source of Variations	Degree of Freedom	Sum of Squares	Mean Squares	F-Value	p-Value
Model	8	3855.630	481.954	3691.64	0.000
Linear	4	104.024	26.006	199.20	0.000
Square	1	154.417	154.417	1182.80	0.000
Interaction	3	14.491	4.830	37.00	0.000
Residual error	631	82.379	0.131	—	—
Total	639	3938.01	—	—	—

TABLE 7.5C

Model Summary for Dependent Variable (Tensile Strength—Igneous Rock)

R^2	Predicted R^2	Adjusted R^2	Standard Error
0.9791	0.9783	0.9788	0.361321

Multiple regression model to predict Young's modulus of the igneous rock types is:

$$E = -56.0046 - 0.3636 \times A + 2.0739 \times D - 0.0082 \times D^2 + 0.0028 \times A \times D$$

Significance of regression coefficients for estimation of Young's modulus is listed in Table 7.6A, which also shows *t*-value and *p*-value. The final ANOVA table of the reduced quadratic model for Young's modulus is shown in Table 7.6B. This table also represents degrees of freedom (DF), mean square (MS), sum of squares (SS), *F*-value, and *p*-value associated with factors. As seen from Table 7.6C, for igneous rocks, the selected model explains 94.57% of the total variation in the observed Young's modulus tests.

TABLE 7.6A

Significance of Regression Coefficients for the Estimation of Young's Modulus (Igneous Rock)

Model Terms for Young's Modulus	Parameter Estimate (Coefficients)	*t*-Value	*p*-Value
Constant	−56. 0046	−10.690	0.000
A	−0.3636	−5.001	0.000
D	2.0739	20.616	0.000
D^2	−0.0082	−16.916	0.000
AD	0.0028	3.985	0.000

TABLE 7.6B

Analysis of Variance (ANOVA) for the Selected Quadratic Model for the Estimation of Young's Modulus (Igneous Rock)

Source of Variations	Degree of Freedom	Sum of Squares	Mean Squares	*F*-Value	*p*-Value
Model	4	6968.746	1742.186	2784.00	0.000
Linear	2	267.114	133.557	213.42	0.000
Square	1	179.070	179.070	286.15	0.000
Interaction	1	9.936	9.936	15.88	0.000
Residual error	635	397.373	0.626	—	—
Total	639	7366.12	—	—	—

TABLE 7.6C

Model Summary for Dependent Variable (Young's Modulus—Igneous Rock)

R^2	Predicted R^2	Adjusted R^2	Standard Error
0.9461	0.9453	0.9457	0.791065

Multiple regression model to predict percentage porosity of the igneous rock types is:

$$n = 33.1804 + 0.0943 \times A - 0.5338 \times D + 0.0022 \times D^2 - 0.0007 \times A \times D$$

Significance of regression coefficients for estimation of percentage porosity is listed in Table 7.7A, which also shows t value and p value. The final ANOVA table of the reduced quadratic model for dry percentage porosity is shown in Table 7.7B. This table also represents degrees of freedom (DF), mean square (MS), sum of squares (SS), F-value, and p-value associated with factors. As seen from Table 7.7C, for igneous rocks, the selected model explains 94.19% of the total variation in the observed percentage porosity tests.

TABLE 7.7A

Significance of Regression Coefficients for the Estimation of % Porosity (Igneous Rock)

Model Terms for % Porosity	Parameter Estimate (Coefficients)	t-Value	p-Value
Constant	33.1804	26.815	0.000
A	0.0943	5.493	0.000
D	−0.5338	−22.467	0.000
D^2	0.0022	18.923	0.000
AD	−0.0007	−4.513	0.000

TABLE 7.7B

Analysis of Variance (ANOVA) for the Selected Quadratic Model for the Estimation of % Porosity (Igneous Rock)

Source of Variations	Degree of Freedom	Sum of Squares	Mean Squares	f-Value	p-Value
Model	4	361.8989	90.4747	2591.90	0.000
Linear	2	17.6990	8.8495	253.52	0.000
Square	1	12.4997	12.4997	358.09	0.000
Interaction	1	0.7109	0.7109	20.37	0.000
Residual error	635	22.1657	0.0349	—	—
Total	639	384.065	—	—	—

TABLE 7.7C

Model Summary for Dependent Variable (% Porosity—Igneous Rock)

R^2	Predicted R^2	Adjusted R^2	Standard Error
0.9423	0.9415	0.9419	0.186833

7.5 Performance Prediction of Derived Models

In fact, the coefficient of correlation between the measured and predicted values is a good indicator to check the prediction performance of the model. However, in this work, Values Account For (VAF; Equation 7.4) and Root Mean Square Error (RMSE; Equation 7.5) indices were calculated to compare the performance of the prediction capacity of predictive models developed (Alvarez and Babuska, 1999; Finol et al., 2001; Gokceoglu, 2002; Yilmaz and Yuksek, 2008; Yilmaz and Yuksek, 2009; Yilmaz and Kaynar, 2011).

$$VAF = \left[1 - \frac{\mathrm{var}(y - y')}{\mathrm{var}(y)}\right] \times 100 \tag{7.4}$$

$$RMSE = \sqrt{\frac{1}{N}\sum_{i=1}^{N}(y - y')^2} \tag{7.5}$$

where y and y' are the measured and predicted values, respectively. If the VAF is 100 and RMSE is 0, then the model will be excellent. Mean absolute percentage error (MAPE), which is a measure of accuracy in a fitted series value, was also used to check the prediction performances of the models (Gokceoglu, 2002; Yilmaz and Kaynar, 2011). MAPE usually expresses accuracy as a percentage (Equation 7.6).

$$MAPE = \frac{1}{N}\sum_{i=1}^{N}\left|\frac{A_i - P_i}{A_i}\right| \times 100 \tag{7.6}$$

where A_i is the actual value and P_i is the predicted value. Lower values of MAPE indicate that there will be a better correlation between predicted values and experimental results.

Using the developed regression models for igneous rocks, performance prediction indices for training as well as test data were calculated and are given in Table 7.8. From the table it is observed that, for igneous rocks, developed model for predicting dry density is less efficient when compared to all other models, as it has low VAF value. MAPE values for test data are 13.354, 15.832, 22.263, 16.669, 8.741, 11.393, and 7.507 for UCS, SRN, dry density, P-wave velocity, TS, E, and percentage porosity, respectively, which indirectly explains the reliability of the predicted models of igneous rocks.

The various regression models developed for metamorphoic and sedimentary rocks are given in Appendix I and Appendix II, respectively.

TABLE 7.8

Performance Prediction Indices of Regression Models (Igneous Rock)

Models		UCS	SRN	Density	P-Wave Velocity	Tensile Strength	Young's Modulus	% Porosity
Training data	VAF	97.453	98.091	77.534	91.023	97.882	94.573	94.192
	RMSE	6.5917	7.1710	0.4372	0.9261	0.4572	8.5119	0.5141
	MAPE	10.971	14.597	15.766	15.497	5.908	11.133	6.701
Test data	VAF	95.448	97.356	63.323	90.953	96.792	93.319	91.123
	RMSE	8.3415	8.0334	0.6132	0.9926	0.6958	8.6372	0.5387
	MAPE	13.354	15.832	22.263	16.669	8.741	11.393	7.507

References

Alvarez, G.M., and Babuska, R. (1999). "Fuzzy model for the prediction of unconfined compressive strength of rock samples." *Int. J. Rock Mech. Min. Sci.*, 36, 339–349.

Bowerman, B.L., and O'Connell, R.T. (1990). *Linear Models: An Applied Approach,* 2nd ed., Boston, Massachusetts: PW-Kent Publishing Company.

Finol, J., Guo, Y.K., and Jing, X.D. (2001). "A rule based fuzzy model for the prediction of petrophysical rock parameters." *J. Pet. Sci. Eng.*, 29, 97–113.

Gokceoglu, C. (2002). "A fuzzy triangular chart to predict the uniaxial compressive strength of the Ankara Agglomerates from their petrographic composition." *Eng. Geol.*, 66, 39–51.

Levin, R.I., and Rubin, D.S. (2006). *Statistics for Management*, 7th ed. New Delhi, Prentice Hall of India.

Yilmaz, I., and Kaynar, O. (2011). "Multiple regression, ANN (RBF, MLP) and ANFIS models for prediction of swell potential of clay soils." *Expert Syst. Appl.*, 38, 5958–5966.

Yilmaz, I., and Yuksek, A.G. (2008). "An example of artificial neural network application for indirect estimation of rock parameters." *Rock Mech. Rock Eng.*, 41(5), 781–795.

Yilmaz, I., and Yuksek, A.G. (2009). "Prediction of the strength and elasticity modulus of gypsum using multiple regression, ANN, ANFIS models and their comparison." *Int. J. Rock Mech. Min. Sci.*, 46(4), 803–810.

8

Application of Artificial Neural Networks

8.1 Introduction

Though statistical relations are good enough to predict the properties fairly well, they are limited by the degree of nonlinearity they can model. Moreover, statistical relations constrain the data along a particular geometry which may not always be favorable to capture the nonlinear relations existing between various parameters. Further, in engineering domains, algebraic and differential equations are used to describe the behavior and functional properties of real systems and to create mathematical models to represent them. Such approaches require accurate knowledge of system dynamics and use of estimation techniques and numerical calculations to emulate the system operation. The complexity of the problem itself may introduce uncertainties that make the modeling nonrealistic or inaccurate. The sound produced during rock drilling is so complicated that accurate modeling will be difficult and may be only approximate it well sometimes, because of the complexity of the nature of the rock drilling process. Artificial neural networks (ANN) have been found to be very efficient in handling nonlinear relationships and intelligent prediction of the required parameters. ANNs implement algorithms that attempt to achieve a neurological-related performance such as learning from experience, making generalization from similar situations, and judging states where poor results are achieved in the past.

A neural network is a massively parallel-distributed processor made up of simple processing units, which has a natural propensity for storing experiential knowledge and making it available for use. One of the advantages is their ability to model complex nonlinear relationships between multiple input variables and requires outputs. In the present work, as the complex interactions between different variables are not completely understood; the ANN approach for modeling is best suited.

ANN resembles a biological brain (Figure 8.1). In the brain there is a flow of coded information from the synapses toward the axon. The axon of each neuron transmits information to a number of other neurons. The neuron receives the information at the synapses from a large number of

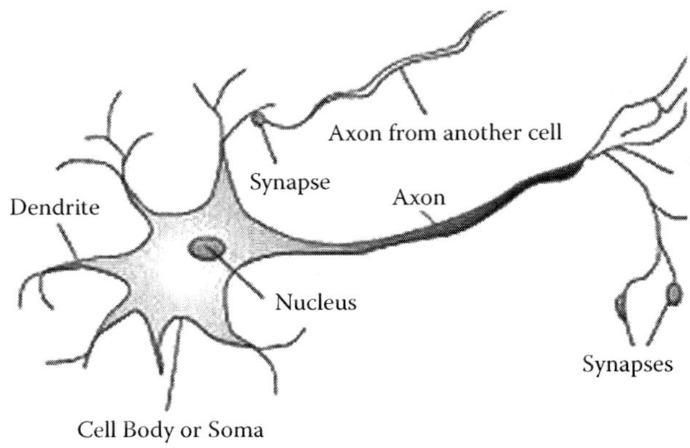

FIGURE 8.1
A biological neuron.

other neurons. Groups of neurons are organized into subsystems and the integration of the following subsystems form the brain:

Dendrites: Accept inputs

Soma: Process the inputs

Axon: Turn the processed inputs to outputs

Synapses: The electro-mechanical contact between the neurons

Figure 8.2 shows the simplified model of an artificial neuron, which may be used to simulate some important aspects of a real biological neuron. An ANN is a group of interconnected artificial neurons, interacting with one another in a concerted manner. In such a system, excitation is applied to the input of the network. Following some suitable operation, it results in a desired response. At the synapses, there is an accumulation of some potential, which in case of artificial neurons is modeled as connection weights. These weights are continuously modified based on suitable learning rules.

Figure 8.3 shows how information is processed through a single neuron. Each node in a layer (except the ones in the input layer) provides threshold of a single value by summing up their input value p_i with the corresponding weight value w_i. Then the neuron's input value n is formed by adding up weighted value (sum), with the bias term b. The bias is added to shift the sum relative to the origin. The net input value then goes into the transfer function f, which produces the neuron output 'a'.

$$a = f\left(\sum_{i=1}^{i=r} w_i p_i + b\right)$$

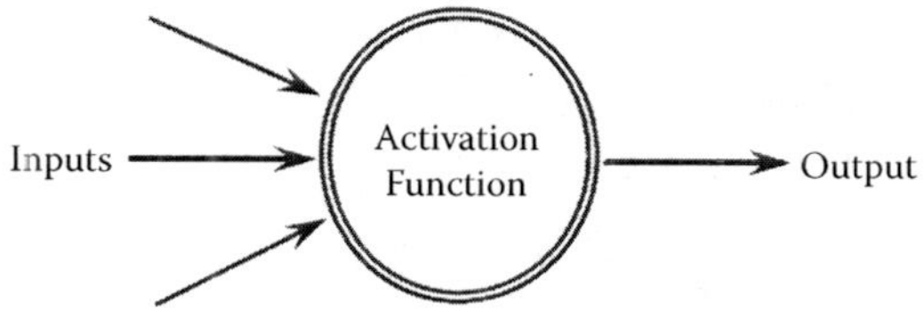

FIGURE 8.2
ANN neuron connection.

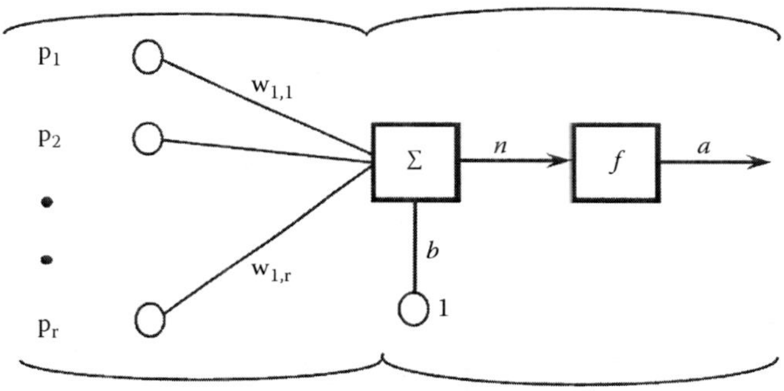

FIGURE 8.3
Information processing through a neuron.

The transfer function f that transforms the weighted inputs into the output "*a*" is usually a nonlinear function, either sigmoid or logistic, which restricts the nodes out between 0 and 1.

ANN is made of large number of neurons and a huge number of interactions between them. According to the structure of the connections, they have been identified as feed forward and recurrent networks. Feed forward networks have one-way connections, from input to the output layer. They are most commonly used for prediction and nonlinear function fitting. Here, the neurons are arranged in the form of layers. Neurons in one layer get inputs from the previous layer and feed their outputs to the next layer. The last layer is called the *output layer*. Layers between the input and output layers are called *hidden layers* and are termed *multilayered networks*.

Figure 8.4 shows a schematic representation of a multilayered feed forward network. The number of nodes within the input and output layers are dictated by the nature of the problem to be solved and the number of input and output variables needed to define the problem. The number of hidden layers and neurons in the hidden layer is usually defined by trial and error method.

ANN studies input output relationships by suitably adjusting the synaptic weights in a process known as training. The weights of the given

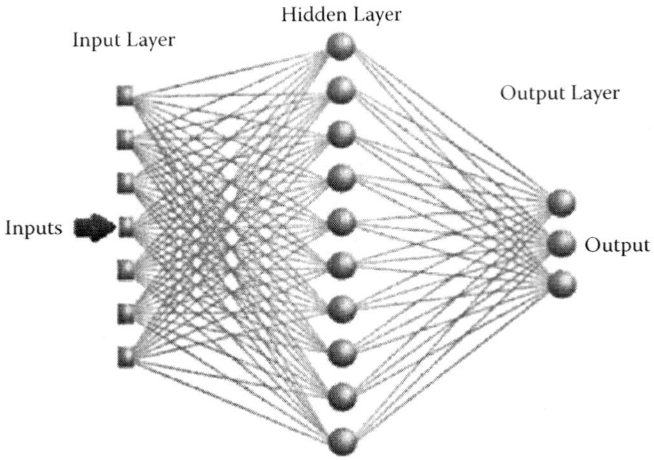

FIGURE 8.4
Multilayered ANN structure.

interconnection are adjusted by means of some learning algorithms. A learning cycle in an ANN model is as shown in Figure 8.5 (Magali and Paul 2003).

Learning methods in neural networks can be broadly classified into two types. They are

1. Supervised learning
2. Unsupervised learning

In supervised learning, target values or desired responses are known and are given to ANN during training so that ANN can adjust its weights to try to match its output to the target values. Before the learning algorithms are applied to update the weights, all the weights are initialized randomly (Haykin 1999). The network using this set of inputs produces its own outputs. These are compared with the target outputs and the difference between them, called the error, is used for modifying the weights.

In the unsupervised learning method (also known as *self-organized learning*), the network develops its classification rules by extracting information from the inputs presented to the network. In other words, using the correlation of input vectors the learning rule changes the network weights to group the input vectors into clusters. By doing so, similar input vectors will produce similar network outputs since they belong to the same cluster.

In this chapter, two types of feed-forward networks, namely, Multilayer Perceptron (MLP) and Radial Basis Function (RBF). These have been used to develop the prediction model of rock properties.

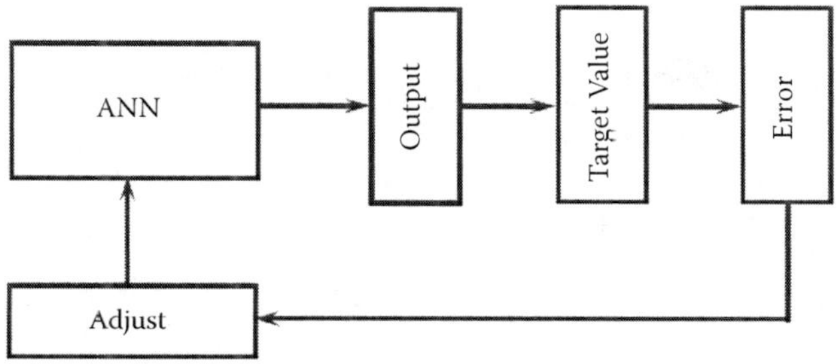

FIGURE 8.5
A learning cycle in an ANN model.

8.2 Multilayer Perceptron

This is one of the popular network architectures used for function approximation, classification, and prediction problems (Haykin, 1999). It is a powerful neural network type, capable of modeling complex relationships between the variables. The architecture of MLP is a multilayered feed-forward neural network, in which nonlinear elements (neurons) are arranged in successive layers and the information flows unidirectionally, that is from the input layer to the output layer through hidden layers. Figure 8.6 shows a typical MLP architecture with the following characteristics:

1. The model of each neuron in the network includes nonlinearity at the output. The nonlinearity used is smooth (i.e., differentiable everywhere). A commonly used form of nonlinearity that satisfies this requirement is the sigmoidal nonlinearity defined by the logistic function: $y_j = \frac{1}{\left(1+e^{(-V_j)}\right)}$ where V_j is the net internal activity of neuron j and y_j is the output of the neuron.
2. The network contains one or more layers of hidden neurons that are not part of input or output of the network. These hidden neurons enable the network to learn complex tasks by extracting progressively more meaningful features from the input patterns.
3. The network exhibits a high degree of connectivity, determined by the synapses of the network.

MLP is trained by using one of the supervised algorithms of which the best known is the *back-propagation algorithm*. The development of this

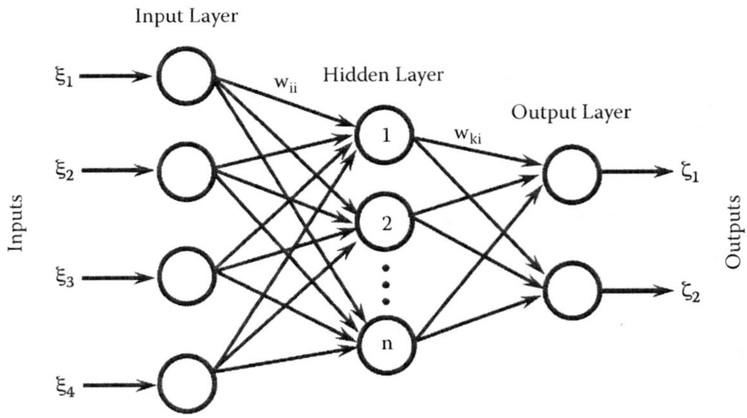

FIGURE 8.6
Generalized MLP architecture.

algorithm is considered a landmark in neural networks, in that it provides a computationally efficient method for training MLPs (Anderson, 1995).

8.3 Back-Propagation (BP) Algorithm

The back-propagation (BP) algorithm allows experimental acquisition of input/output mapping knowledge within multilayer networks. There are basically two passes through the different layers of the network: a feed-forward pass and a backward pass. In the forward pass, an input pattern is submitted and propagated through the network, layer by layer. A set of outputs is produced as the actual response of the network. During the forward pass, the synaptic weights are all fixed, and in the backward pass, the synaptic weights are all adjusted, depending on the error between the actual output and the target output. The process is continued until all the input patterns from the training set are learned with an acceptable overall error. The error is cumulative and computed over the entire training set. This computation is called the *training epoch*. During the testing phase, the trained network itself operates in a feed-forward manner (Haykin 1999). The BP algorithm is presented below.

1. Initialize the weights and biases to small random values.
2. Chose an input pattern $\left\{\xi_i^\mu, \xi_k^\mu\right\}$ from the training set and present ξ_i^μ to the input layer where $\mu = 1, 2, 3, \ldots, n$ are the number of patterns

and i = 1, 2, . . . are the number of features and k = 1, 2, 3, . . . are the number of output features.

3. Compute the activation of the neurons in the hidden layer $v_j^\mu = \frac{1}{\left(1+e^{-\left(\sum w_{ji}\xi_i\right)}\right)}$

4. Compute the output of each neuron in the output layer $O_k^\mu = \frac{1}{\left(1+e^{-\left(\sum w_{kj}v_j\right)}\right)}$

5. Compute the mean squared error (MSE) $E^\mu = \frac{1}{2}\sum_\mu \left(\xi_k^\mu - O_k^\mu\right)^2$

6. If $E^\mu < E_{min}$, go to step 8.

7. Compute $\delta_k^\mu = \left(\xi_k^\mu - O_k^\mu\right)O_k^\mu\left(1-O_k^\mu\right)$

   Update the weights between the outputs and the hidden layer.

   $w_{kj}^{new} = w_{kj}^{old} + \Delta w_{kj}$, where $\Delta w_{kj} = \alpha\eta\sum_\mu \delta_k^j v_j^\mu$, where η is the learning rate and α is the momentum coefficient.

   Update the weights between the hidden and input layer

   $w_{ji}^{new} = w_{ji}^{old} + \Delta w_{ji}$, where $\Delta w_{ji} = \alpha\eta\sum_\mu \delta_j^\mu \xi_i^\mu$

   Go to step 2.

8. Save all the weights and exit.

The performance of the back-propagation algorithm depends on the initial setting of weights, learning rate, output functions of the units, presentation of the training data, and the specific pattern recognition task such as classification, prediction, or mapping:

1. *Initial weights*—The network weights are initialized to small random values. The initialization strongly affects the final solution.
2. *Transfer function of the Nodes*—For calculating the value of δ in the backward pass, the requirement is that the activation function should be differentiable. One of the most widely used functions, which is continuously differentiable and nonlinear is the sigmoidal nonlinearity. A particular form defined for the sigmoidal nonlinearity is given by $f(x) = \frac{1}{1+e^{-x}}$, where x is the net internal activity of the neuron and $0 \le f(x) \le 1$. This has been used both for the nodes in the hidden layer and output layer.

3. *Learning rate*—The effectiveness and convergence of back propagation algorithm depends significantly on the value of the learning rate η. By trial and error, the value for the learning rate has to be selected, which provides an optimum solution. The value is generally lesser than 1.
4. *Momentum coefficient*—The momentum term is generally used to accelerate the convergence of the error BP algorithm. This involves the use of momentum coefficient α. This is a simple method of increasing the rate of learning and yet avoids the danger of instability. The value chosen is generally lesser than 1.
5. *Number of hidden neurons*—The size of hidden layer is one of the most important considerations, while solving actual problems using MLP neural network. The optimal number of hidden nodes in any network for solving any given problem is determined by trial and error. Hidden units play a critical role in the operation of multilayer perceptron with BP algorithm learning as they act as feature detectors.

In this work, the ANN model was developed by using MatLab R2008 software for windows. Standard training functions used in the neural network toolbox were used for the above purpose. Four different fast-training algorithms have been used for training the MLP network. Gradient descent with momentum and adaptive learning backpropagation algorithm (*traingdx*) is an algorithm that uses variable learning rate for training. Initially, the network outputs and error are calculated. At each epoch, new weights and biases are calculated by using the current learning rate, and if this error exceeds the older error by more than the predefined ratio, new weights and biases are discarded. In addition the learning rate is decreased. If the new error is less than the old error, the new weights and biases are kept and the learning rate is increased. Hence, in this way an optimal learning rate is in the local terrain (Mathworks, 2008).

The resilient back-propagation algorithm (*trainrp*) eliminates the harmful effects of partial derivatives. The only sign of the derivative is used to determine the direction of the weight updates. The magnitude of the derivative has no effect on the weight update. The update value for each weight and bias is increased by a predefined value whenever derivative of the performance function with respect to that weight has the same sign for the two successive iterations. The update value is decreased by that value whenever the derivative with respect to that weight changes signs from the previous iteration (Mathworks 2008).

With the *scaled conjugate gradient algorithm* (*trainscg*), a search is performed along the conjugate directions, which generally produces faster convergence than the steepest descent algorithms (Mathworks, 2008).

The *Levenberg–Marquardt algorithm* (*trainlm*), was designed to approach second-order training speed without having to compute the Hessian matrix. When the performance function has the from of a sum of squares (as is typical in training feed-forward networks), then the Hessian matrix can be approximated as $H = J^TJ$ and the gradient can be computed as $g = J^Te$, where J is the Jacobian matrix that contains first derivatives of the network errors with respect to the weights and biases and *e* is a vector of network errors. The Levenberg–Marquardt algorithm uses this approximation to the Hessian matrix in the following Newton-like update.

$$X_{k+1} = X_k - [J^TJ + \mu I]^{-1} J^Te$$

When the scalar μ is zero, this is just Newton's method, using the approximate Hessian matrix. When μ is large, this becomes gradient descent with a small step size. In the Levenberg–Marquardt method, μ is decreased after each successful step (reduction in performance function) and is increased only when a tentative step would increase the performance function. In this way, the performance function is always reduced at each iteration of the algorithm (Mathworks, 2008).

8.4 Multilayer Perceptron (MLP) Model

The ANN model based on MLP was developed by using experimental data. Different training algorithms were compared in order to develop a best possible MLP architecture based on good generalization ability and a compact structure. In this chapter, a detailed procedure used for developing the optimized MLP model for igneous rock category is given. A similar procedure was employed and results were obtained for other categories of rocks.

For MLP models of igneous rocks, similar to multiple regression modeling, approximately 70% (640 patterns) of the experimental data were used in the training set, and the remaining 256 patterns (approximately 30%) were employed for testing of the developed models. Drill bit diameter, bit speed, penetration rate, and an A-weighted equivalent sound level were used as the input parameters. These input parameters cover the entire problem domain under study and are effective in their prediction. Rock properties such as UCS, SRN, dry density, P-wave velocity, tensile strength, Young's modulus, and percentage porosity were the output parameters for the model. To ensure that each input provides an equal contribution in the ANN, the inputs to the model were preprocessed and scaled into a common numeric range (0, 1). A network with one hidden layer was used with a tan sigmoid activation function in the hidden layer and output layer. A mean square error (MSE) of 10^{-4}, a minimum gradient of 10^{-5}, and maximum number of epochs of 2000

TABLE 8.1

Performances of Different Training Algorithms

Sl. No.	Training Algorithm	Network Architecture	Number of Epochs	Time Taken for Convergence (sec)	Mean Square Error
1	traingdx	4 : 10 : 7	1997	128.06	0.00028517
2	trainrp	4 : 15 : 7	2000	136.25	0.00010124
3	trainscg	4 : 20 : 7	768	111.84	0.00010001
4	trainlm	4 : 45 : 7	21	54.26	0.00010000

were used. The training process would stop if any of these conditions were met. The initial weights and biases of the network were fixed randomly.

The number of neurons in the hidden layer has a significant influence on the performance of MLP. For each of the training algorithm, a number of trials were conducted initially to fix the number of neurons in the hidden layer. The number of neurons, for which MSE is minimum, was selected as the optimum number of neurons in the hidden layer, as there is no standard procedure finding the optimum number of neurons in the hidden layer. The architecture and performance of the network using different training algorithms are given in Table 8.1. From the table it is clear that *trainlm* converges faster than all other training algorithms as the number of epochs as well as the time taken for convergence is less. Similar observations were made for other categories of rocks.

The variation of MSE with the number of neurons in the hidden layer for *trainlm* algorithm is shown in Figure 8.7. From the figure it was observed that for the performance model when the number of neurons in the hidden layer was 5, the error was 0.00015528, and it decreased as the number of neurons increased and reached minimum of 0.0001 for 45 and on, increasing the number of neurons further, MSE gradually increased. The final architecture of the MLP model is shown in Figure 8.8, where n = 45.

The trained network was tested for its performance as in the case of multiple regression model. Table 8.2 shows the performance prediction indices of different outputs for the training and test data of igneous rocks using different training algorithms.

From the above table, it is clear that, VAF values are maximum, RMSE and MAPE values are minimum for the network using *trianlm* algorithm when compared to other models for both training and testing data, thereby establishing the superiority of the *trainlm* algorithm.

8.5 Radial Basis Function (RBF) Neural Network

There are different forms of designing a supervised neural network. The back-propagation algorithm for the design of multilayer perceptron (under supervision) can be viewed as an application of an optimization

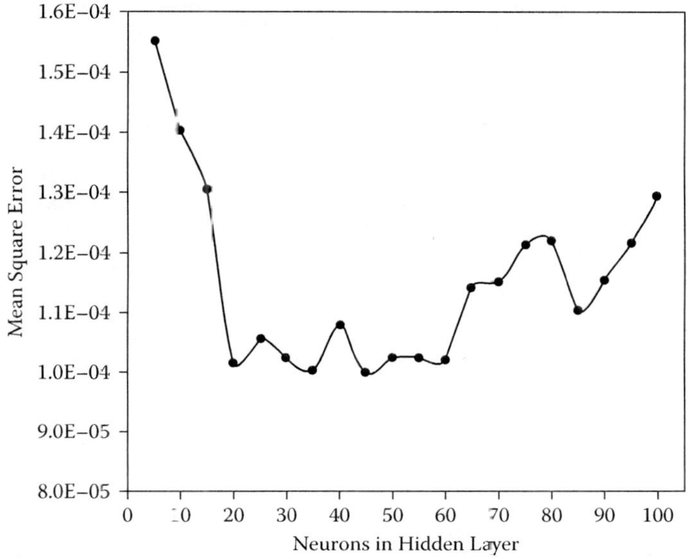

FIGURE 8.7
Variation of MSE with number of neurons in the hidden layer.

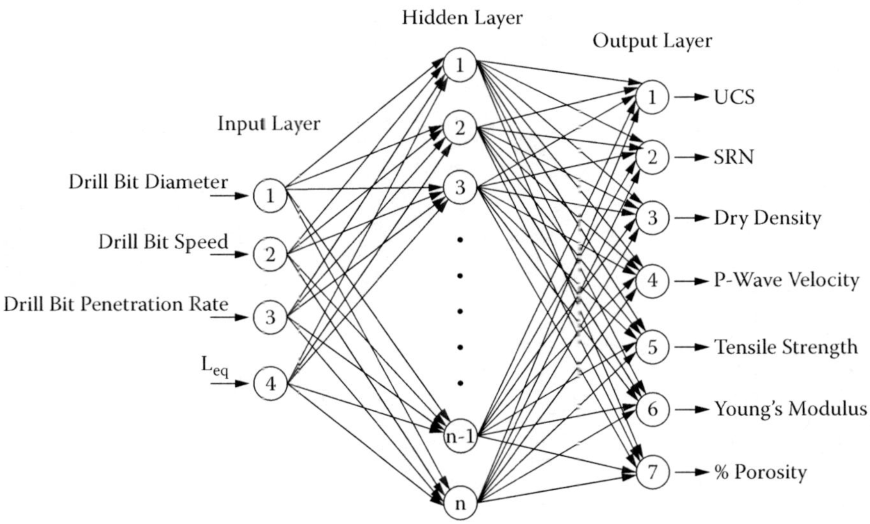

FIGURE 8.8
Architecture of developed MLP model.

TABLE 8.2

Performance Prediction Indices of Different Training Algorithms

Training Algorithms			UCS	SRN	Density	P-Wave Velocity	Tensile Strength	Young's Modulus	% Porosity
traingdx	Training data	VAF	69.572	69.705	62.504	65.594	77.994	77.224	74.561
		RMSE	11.5229	9.7203	9.3618	5.8622	6.4063	19.1825	10.4233
		MAPE	19.575	29.790	25.543	17.896	14.473	18.391	20.599
	Test data	VAF	66.233	67.643	58.636	61.943	77.284	74.654	71.881
		RMSE	16.8730	10.8704	10.4880	6.9319	8.9395	21.7948	10.9418
		MAPE	22.072	31.876	29.914	18.450	17.908	20.803	22.448
trainrp	Training data	VAF	76.538	71.076	65.555	77.498	82.3350	82.293	77.912
		RMSE	10.5161	8.6863	8.6235	4.6192	5.9189	19.1408	9.3641
		MAPE	18.910	28.795	20.477	16.863	13.564	17.354	18.869
	Test data	VAF	74.700	70.349	62.006	77.065	77.512	82.099	76.900
		RMSE	16.8926	9.3866	8.7576	5.8772	7.3759	20.3721	10.5852
		MAPE	19.405	28.966	21.524	17.263	17.496	19.428	16.560
trainscg	Training data	VAF	79.932	76.894	65.525	79.702	88.915	86.567	78.138
		RMSE	4.7129	7.6430	8.0557	4.1087	5.1246	14.0497	8.2238
		MAPE	18.530	27.433	19.962	12.212	12.524	15.381	12.417
	Test data	VAF	76.623	69.813	64.971	70.080	84.360	84.999	72.496
		RMSE	6.2361	8.8356	8.1522	5.1506	6.9136	15.0848	10.0032
		MAPE	24.329	28.457	20.773	13.831	16.285	18.672	14.218
trainlm	Training data	VAF	**98.889**	**98.838**	**92.329**	**96.563**	**98.748**	**96.626**	**95.658**
		RMSE	**6.4410**	**3.7701**	**0.2511**	**0.6260**	**0.4244**	**6.2074**	**0.4530**
		MAPE	**8.027**	**6.601**	**9.233**	**13.254**	**5.139**	**10.220**	**3.0355**
	Test data	VAF	**96.051**	**98.763**	**84.132**	**91.347**	**97.914**	**95.208**	**94.710**
		RMSE	**7.5547**	**6.4733**	**0.4681**	**0.7182**	**0.5649**	**7.0452**	**0.4730**
		MAPE	**9.512**	**10.087**	**17.068**	**15.371**	**7.016**	**10.432**	**7.2190**

method known in statistics as *stochastic approximation*. Another approach can be viewing the design of a neural network as a curve-fitting problem in high-dimensional space. This involves finding a surface in a multidimensional space that provides the best fit to the training data, with the criterion for "best fit" being measured in some statistical sense. Correspondingly, generalization is equivalent to the use of this multidimensional surface to interpolate the test data (Anderson 1995). This is indeed the motivation behind the method of radial basis functions. In the context of a neural network, the hidden units provide a set of "functions" that constitute an arbitrary "basis" for the input patterns (vectors) when they are expanded into the hidden unit space. Therefore, these functions are called "radial basis functions."

The main advantages claimed for the RBF model are its simplicity and the ease of implementation. The learning and generalization ability of these networks are excellent. The construction of a radial basis function network in its most basic form involves three entirely different layers. The input layer is made up of source nodes (sensory units). The second layer is a hidden layer which serves a different purpose from that in a multilayer perceptron. The output layer supplies the response of the network to the activation patterns applied to the input layer. The transformation from the input space to the hidden unit space is nonlinear, whereas from the hidden unit space to the output space it is linear. Figure 8.9 shows the typical RBF architecture.

The RBF network is a single hidden layer feed-forward neural network. Each node has two parameters, a center c_j and a width σ_j. This center is used to compare with the network input vector to produce a radially symmetrical response. The width controls the smoothness properties of the interpolating function. Response of the hidden layer are scaled by the connection weights of the output layer and then combined to produce the network output.

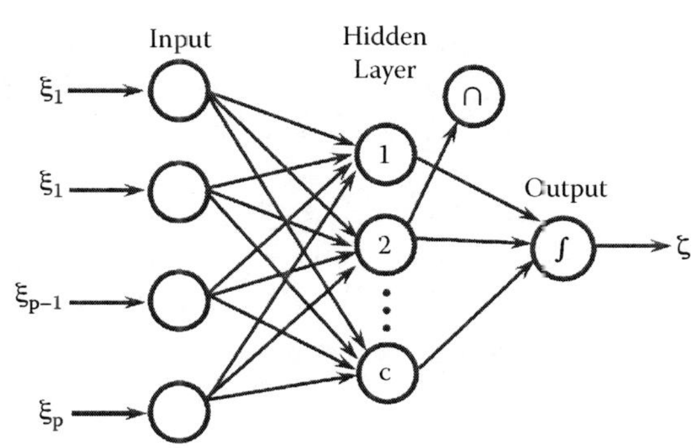

FIGURE 8.9
General architecture of RBF network.

RBFs have been shown to have universal approximation capability. In the classical approach to RBF network implementation, the basic functions are usually chosen as Gaussian, and the number of hidden units is fixed based on some properties of the input data. The weights connecting the hidden and output units are estimated by least squares methods. The disadvantage with the classical approach is that it is not suitable for sequential learning, and it also results usually in too many hidden units. The RBF network requires less computation time for learning and produces a more compact topology than other neural networks (Haykin, 1999).

The learning process undertaken by a RBF network may be visualized as follows. The linear weights associated with the output units of the network tend to evolve on a different time scale compared to the nonlinear activation function of the hidden units. Thus, as the hidden layer's activation functions evolve slowly in accordance with some nonlinear optimization strategy, the output layer's weights adjust themselves rapidly through a linear optimization strategy. Since different layers of an RBF network perform different tasks, it is reasonable to separate the optimization of the hidden and output layers of the network by using different techniques and perhaps operating on different time scales (Lowe, 1991). There are different learning strategies available for the design of an RBF network. The performance of an RBF network depends on the number and positions of the RBFs in the hidden layer, their shape, and the method used for learning the input–output mapping. In this study, the input layer consists of four nodes. The RBF units are defined by two parameters—center and width of the RBF unit. The output V_j of a jth RBF unit for an input pattern x is given by

$$V_j = \exp\left(\frac{\left\|x - c_j\right\|^2}{2\sigma_j^2}\right) \tag{8.1}$$

where c_j and σ_j are, respectively, the center and width of the jth RBF unit, and $\|\ \|$ is the Euclidean distance.

The output nodes are nonlinear with sigmoid activation function. Seven nodes are used at the output layer to denote different rock properties. The learning strategies for RBF considered in this study can be classified as

1. Strategies based on random selection of RBF centers from the training data.
2. Strategies based on unsupervised selection of RBF centers.

The performance of the network for different strategies adopted for changing the centers of the hidden units are explored in this study. In the first strategy, centers were fixed and selected at random. This approach does produce reasonable results, when the training data are distributed in a representative manner. In the second strategy, Cluster-Dependent Weighted Fuzzy C Means (CDWFCM) clustering algorithms were used to initialize the centers of the RBF units. The CDWFCM strategy provided improved results when compared to fixed centers. In both the strategies, the widths of the RBF units were evaluated using

$$\sigma_j = \frac{d_{max}}{\sqrt{2m}} \tag{8.2}$$

where d_{max} is maximum Euclidean distance between chosen centers and m is the number of centers. This formula ensures that the individual RBFs are not too peaked or too flat; both these extreme conditions should be avoided. As an alternative to Equation 8.2, one may use individually scaled centers with broader widths in areas of lower data density, which require experimentation with training data (Haykin 1999). However, in another approach, the width was fixed at a constant value.

In this work, a comparative evaluation was performed between the two learning strategies, and two approaches were used to determine the widths of the basis functions for prediction of rock properties. The learning rate parameter η and momentum coefficient α were maintained constant at 0.8 and 0.4, respectively. The upper limit on the learning cycles was kept at 2000 epochs to observe the network convergence behavior, and the network goal was fixed at 10^{-4} for all the studies.

8.5.1 Radial Basis Function Neural Network (RBFNN) with Fixed Centers Selected at Random

This is the simplest approach in which the locations of the centers of the hidden layer may be chosen randomly from the training data set. The activation functions of the hidden units are Gaussian functions given by Equation 8.1, and widths are computed using Equation 8.2. The weights in the output layer of the network are estimated by the gradient descent method. The algorithm is described below:

1. Select the number of RBF units (c) arbitrarily and minimum error $E_{min} = 0.0001$.
2. Initialize their centers from training pattern randomly and $E_{tot} = 0$, iter = 0.
3. iter = iter + 1, choose the input–output pair (ξ_i, ζ_k).
4. Compute the hidden layer output V_j, using Equation 7.1.
5. Compute output using $O_k = \frac{1}{\left(1+e^{-\sum w_{kj}V_j}\right)}$, where k is the number of units in the output layer (k = 7).

6. Compute the error $E = (O_k - \zeta_{k})*(O_k - \zeta_{k})$ and $E_{tot} = E_{tot} + E$.
7. The change in weights are calculated as follows:

$$\delta_k = (O_k - \zeta_k)*O_k*(1-O_k)$$

$$\Delta w_{kj} = \delta_k * V_j * \eta * \alpha$$

$$w_{kj} = w_{kj} + \Delta w_{kj}$$

8. If $E_{tot} > E_{min}$ & iter <2000, then go to Step 3.
9. Stop.

Table 8.3A shows the network error for the different numbers of selected centers. It was observed that the optimum number of centers was 260, corresponding to minimum training error of 0.000139. Hence, 260 was selected as optimum number of centers. Performance indices of the thusly evolved network for igneous rocks are given in Table 8.3B.

In another study, widths have been kept constant for all the RBF units. The network was trained for different values of widths as shown in Table 8.4A. Variation of network error with fixed values of widths for 260 RBF units is also given in this table. It is clear from the table that the error decreased with the increase in the value of the width. The optimum width value has been chosen as 0.14 since the MSE was least corresponding to this width. Beyond this value of the width, the error started to increase.

TABLE 8.3A

Variation of MSE with number of RBF Units (Random Selection of Centers—Varying Width)

Number of centers	100	150	250	**260**	270	280
MSE × 10^{-3}	0.865	0.637	0.270	**0.139**	0.223	0.472
RBF width	0.1101	0.1369	0.1306	**0.1211**	0.1217	0.1284

TABLE 8.3B

Performance of RBF NN with 260 RBF Units (Random Selection of Centers—Varying Width)

Rock Properties		**UCS**	**SRN**	**Density**	**P-Wave Velocity**	**Tensile Strength**	**Young's Modulus**	**% Porosity**
Training data	VAF	93. 477	89.772	73.654	83.496	84.535	88.933	84.228
	RMSE	11.7261	8.8963	0.8235	1.9892	1.0769	10.0418	1.8964
	MAPE	15.015	17.685	18.347	17.443	10.164	15.378	9.389
Test data	VAF	91.705	85.389	70.866	81.935	82.592	85.839	82.970
	RMSE	17.0226	9.9866	0.9534	2.0771	1.4251	10.9462	2.0322
	MAPE	18.985	19.264	19.122	19.243	10.991	16.024	11.660

TABLE 8.4A

Variation of MSE with Widths for 260 RBF Units (Fixed Width)

RBF Width	0.11	0.12	0.13	0.14	0.15	0.16
Epochs	2000	2000	2000	1781	2000	2000
MSE	0.000497	0.000299	0.000201	0.000103	0.000227	0.000332

TABLE 8.4B

Performance of RBF NN Using Random Selection of Centers with 260 RBF Units and Fixed Width (0.14)

Rock Properties		UCS	SRN	Density	P-Wave Velocity	Tensile Strength	Young's Modulus	% Porosity
Training Data	VAF	97.538	98.176	79.555	95.498	98.3350	95.293	94.912
	RMSE	6.5161	6.6863	0.3235	0.7192	0.4389	7.1408	0.4641
	MAPE	9.910	8.795	13.477	14.863	5.564	10.354	5.869
Test Data	VAF	95.700	97.399	72.006	91.065	97.512	94.099	93.900
	RMSE	7.8926	7.3866	0.5576	0.8772	0.6759	7.3721	0.5152
	MAPE	10.405	14.966	18.524	16.263	7.496	10.428	7.360

The performance indices for the igneous rocks, of the network for the width chosen are shown in Table 8.4B. The results show that the values of RMSE and MAPE slightly decreased, and VAF value slightly increased when compared to the values obtained using varying width method.

8.5.2 RBF Centers Initialized Using Cluster Dependent Weighted Fuzzy C Means (CDWFCM)

In this method, the centers were initialized from the input patterns using the CDWFCM algorithm. The CDWFCM algorithm has been proposed by Hadjahmadi et al. (2008), in which the weights w_k depend on a particular cluster. These weights can change and be updated during the clustering process. The CDWFCM algorithm is given below:

1. Select number of centers c, set fuzziness parameter, m = 2, set minimum error, E_{min} = 0.0001 and iter = 0.
2. Generate an initial set of prototypes $v_0 = \{v_1, v_2, \ldots, v_c\}$.
3. iter = iter + 1,

$$\eta_i = \frac{K \times \sum_{k=1}^{n} \left(u_{ik}^m d_{ik}^2\right)}{\sum_{k=1}^{n} u_{ik}^m} \quad \text{where } d_{ik} = \|x_k - v_i\|; \quad i = 1 \text{ to } c; \quad K > 0, \text{ usually } K = 1$$

$$w_{ik} = \left(\frac{1}{d_{ik} + \varepsilon}\right)^{\frac{2}{(\eta_i - 1)}} \quad \begin{array}{l} \textit{where } i = 1 \textit{ to } c; \quad k = 1 \textit{ to } n \\ \textit{and } \varepsilon > 0 \textit{ is added to denominator to avoid division by zero} \end{array}$$

$$u_{ik} = \left(\sum_{j=1}^{c}\left(\frac{w_{ik}d_{ik}}{w_{jk}d_{jk}}\right)^{\frac{2}{(m-1)}}\right)^{-1} \quad \begin{array}{l} \textit{where } d_{ik} = \|x_k - v_i\|; \quad d_{jk} = \|x_k - v_j\| \\ \textit{and } i = 1 \textit{ to } c; \quad k = 1 \textit{ to } n \end{array}$$

$$v_i = \frac{\sum_{k=1}^{n}\left(w_{ik}u_{ik}^{m}x_k\right)}{\sum_{k=1}^{n} w_{ik}u_{ik}^{m}}; \mathrm{i} = 1 \text{ to c}$$

$$E_{iter} = \|v_i - v_0\|$$

$$\mathrm{v}_0 = \mathrm{v}_\mathrm{i}$$

4. If iter < N (maximum number of iterations) and $E_{iter} > E_{min}$, then go to step 3.

Table 8.5A shows the network error for the different numbers of selected centers with centers initialized using CDWFCM algorithm. It was observed that the optimum number of centers was 170 corresponding to minimum training error of 0.000128. Hence, 170 was selected as optimum number of centers. Performances of the so-evolved network for igneous rocks are given in Table 8.5B.

In another study, widths have been kept constant for all the RBF units, and the network was trained for different values of widths. Table 8.6A shows the variation of network error with fixed values of widths for 170 RBF units. It is clear from the table that the error decreased with the increase in the value of the width. The optimum width value has been chosen as 0.09 since the MSE was least corresponding to this width. Beyond this value of the width, the error started to increase.

The performance for the igneous rocks of the network for this width chosen is shown in Table 8.6B. The results show that the values of RMSE and MAPE

TABLE 8.5A

Variation of MSE With Number of RBF Units (Centers Initialized Using CDWFCM - Varying Width)

Number of centers	100	130	160	170	180	200
MSE x 10^{-3}	0.697	0.495	0.177	0.128	0.183	0.374
RBF width	0.1301	0.1153	0.1006	0.0913	0.1017	0.1244

TABLE 8.5B

Performance of RBF NN With 170 RBF Units (Centers Initialized Using CDWFCM—Varying Width)

Rock Properties		UCS	SRN	Density	P-Wave Velocity	Tensile Strength	Young's Modulus	% Porosity
Training data	VAF	98.167	98.174	94.348	96.629	97.428	97.482	95.717
	RMSE	6.4787	3.0270	0.5147	0.7528	0.4698	5.0278	0.8349
	MAPE	8.159	7.169	6.993	10.692	5.387	9.103	2.662
Test data	VAF	97.296	93.368	90.178	94.289	96.829	93.593	93.549
	RMSE	7.0278	6.6038	0.6875	0.8201	0.5314	7.1673	1.1584
	MAPE	9.259	9.106	10.006	12.931	6.027	9.618	6.891

TABLE 8.6A

Variation of MSE With Widths for 170 RBF Units (Centers Initialized Using CDWFCM—Fixed Width)

RBF width	0.05	0.08	0.09	0.10	0.13	0.15
Epochs	2000	2000	1891	2000	2000	2000
MSE	0.000397	0.000168	0.000109	0.000173	0.000642	0.000723

TABLE 8.6B

Performance of RBF NN Using CDWFCM With 170RBF Units and Fixed Width (0.09)

Rock Properties		UCS	SRN	Density	P-Wave Velocity	Tensile Strength	Young's Modulus	% Porosity
Training data	VAF	99.761	99.798	96.626	98.882	99.982	98.897	97.897
	RMSE	5.7357	2.8340	0.2098	0.5834	0.3644	4.8914	0.4184
	MAPE	7.838	6.310	6.384	10.076	4.896	8.842	2.0969
Test data	VAF	97.692	98.904	92.078	96.175	98.460	96.976	95.332
	RMSE	6.9359	6.4083	0.2759	0.6202	0.4374	6.9553	0.4585
	MAPE	8.752	8.710	9.806	12.531	5.845	9.494	6.197

slightly decreased, and VAF value slightly increased when compared to the values obtained using varying width methods. Hence, RBF NN using CDWFCM performed optimally with the configuration 4:170:7, with a width of 0.09.

8.6 ANN and Regression Models—A Comparison

In this section, the best architecture in each of the ANN methods has been compared with the regression method. The results of comparison have been presented in Table 8.7 for igneous rocks.

TABLE 8.7

Comparison of Performances of the Developed Models (Igneous Rocks)

			UCS	SRN	Density	P-Wave Velocity	Tensile Strength	Young's Modulus	% Porosity
Multiple Regression	Training data	VAF	97.453	98.091	77.534	91.023	97.882	94.573	94.192
		RMSE	6.5917	7.1710	0.4372	0.9261	0.4572	8.5119	0.5141
		MAPE	10.971	14.597	15.766	15.497	5.908	11.133	6.701
	Test data	VAF	95.448	97.356	63.323	90.953	96.792	93.319	91.123
		RMSE	8.3415	8.0334	0.6132	0.9926	0.6958	8.6372	0.5387
		MAPE	13.354	15.832	22.263	16.669	8.741	11.393	7.507
MLP (trainlm) 4:45:7	Training data	VAF	98.889	98.838	92.329	96.563	98.748	96.626	95.658
		RMSE	6.4410	3.7701	0.2511	0.6260	0.4244	6.2074	0.4530
		MAPE	8.027	6.601	9.233	13.254	5.139	10.220	3.0355
	Test data	VAF	96.051	98.763	84.132	91.347	97.914	95.208	94.710
		RMSE	7.5547	6.4733	0.4681	0.7182	0.5649	7.0452	0.4730
		MAPE	9.512	10.087	17.068	15.371	7.016	10.432	7.2190
RBF NN (4:260:7) (Fixed width = 0.14)	Training data	VAF	97.538	98.176	79.555	95.498	98.3350	95.293	94.912
		RMSE	6.5161	6.6863	0.3235	0.7192	0.4389	7.1408	0.4641
		MAPE	9.910	8.795	13.477	14.863	5.564	10.354	5.869
	Test data	VAF	95.700	97.399	72.006	91.065	97.512	94.099	93.900
		RMSE	7.8926	7.3866	0.5576	0.8772	0.6759	7.3721	0.5152
		MAPE	10.405	14.966	18.524	16.263	7.496	10.428	7.360
CDWFCM (4:170:7) (Fixed width = 0.09)	Training data	VAF	99.761	99.798	96.626	98.882	99.982	98.897	97.897
		RMSE	5.7357	2.8340	0.2098	0.5834	0.3644	4.8914	0.4184
		MAPE	7.838	6.310	6.384	10.076	4.896	8.842	2.0969
	Test data	VAF	97.692	98.904	92.078	96.175	98.460	96.976	95.332
		RMSE	6.9359	6.4083	0.2759	0.6202	0.4374	6.9553	0.4585
		MAPE	8.752	8.710	9.806	12.531	5.845	9.494	6.197

A few observations that can be made are given below:

1. The MAPE and RMSE values for all the variables in the case of regression models were higher than the ANN models.
2. The prediction accuracy, that is, VAF values for the regression models were lower when compared to the ANN models for all the variables.
3. Performance indices show that the MLP model is better than RBFNN with fixed centers selected at random for prediction of rock properties.
4. RBF model with RBF centers initialized using the CDWFCM algorithm outperformed the perdition capability of MLP model as well as RBFNN with fixed centers selected at random with more than 90% prediction capability for all the prediction variables. Hence, it can be concluded that RBF NN with centers initialized using CDWFCM can be effectively used for the prediction of rock properties using sound level produced during rotary drilling.

In order to show the deviations from the experimental values of igneous rock properties, the percentage error between the predicted values using the constructed models and the experimental values were also calculated and are shown in Figure 8.10 to Figure 8.16. These figures indicate that the

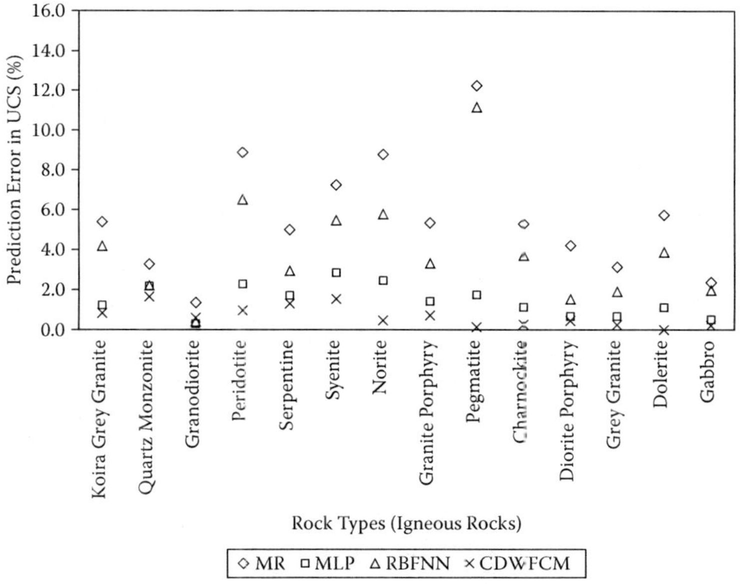

FIGURE 8.10
The error between the predicted values in UCS (%) using the constructed models and the experimental values.

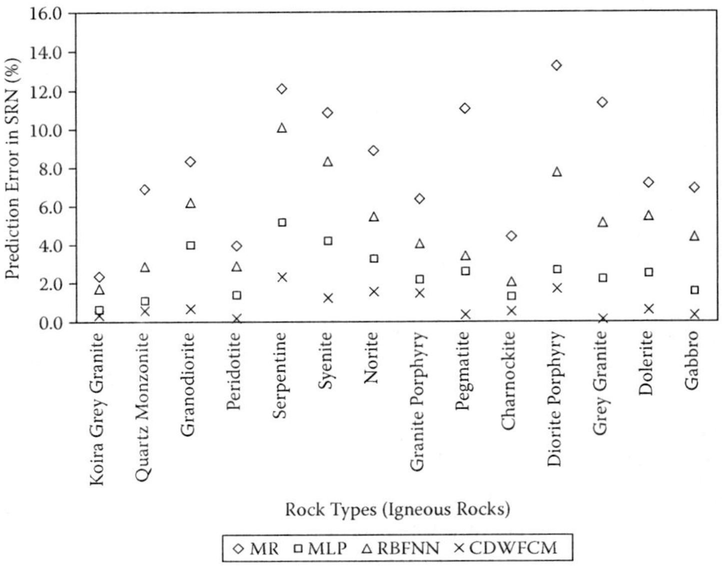

FIGURE 8.11
The error between the predicted values in SRN (%) using the constructed models and the experimental values.

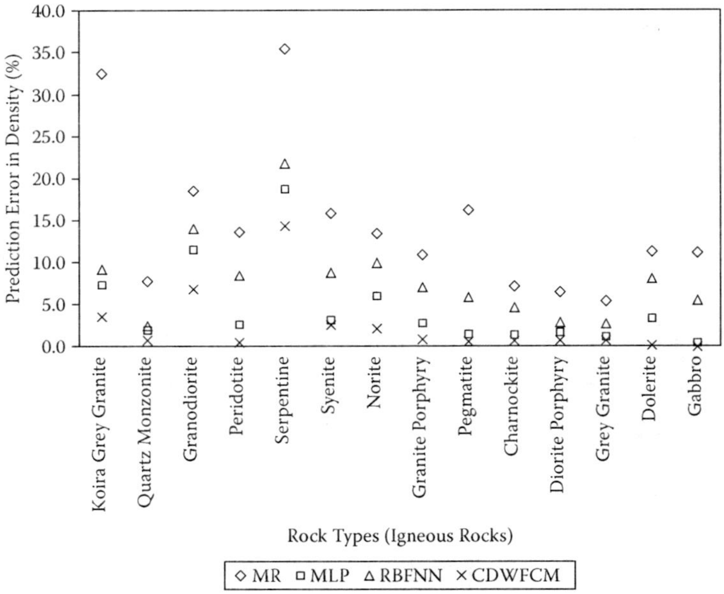

FIGURE 8.12
The error between the predicted values in dry density (%) using the constructed models and the experimental values.

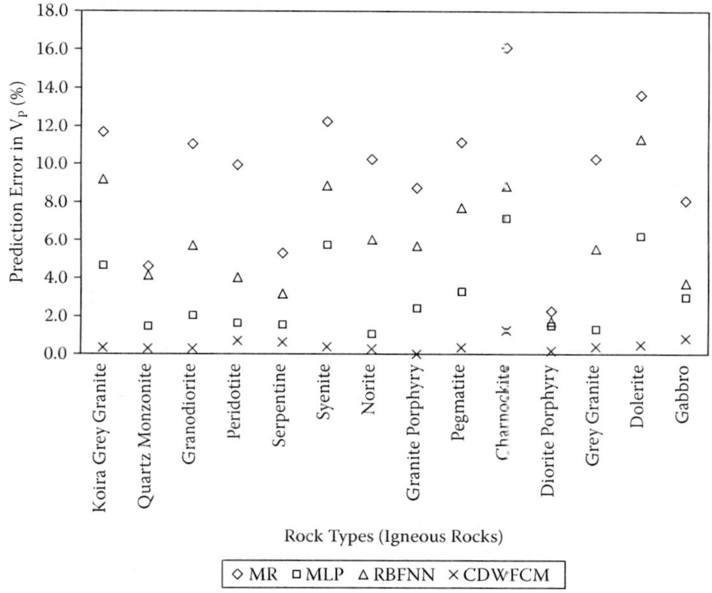

FIGURE 8.13
The error between the predicted values in P-wave velocity (%) using the constructed models and the experimental values.

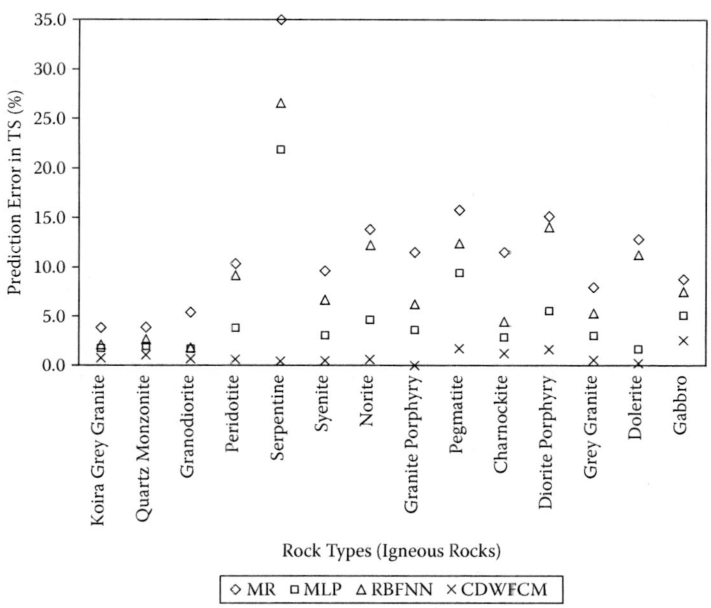

FIGURE 8.14
The error between the predicted values in tensile strength (%) using the constructed models and the experimental values.

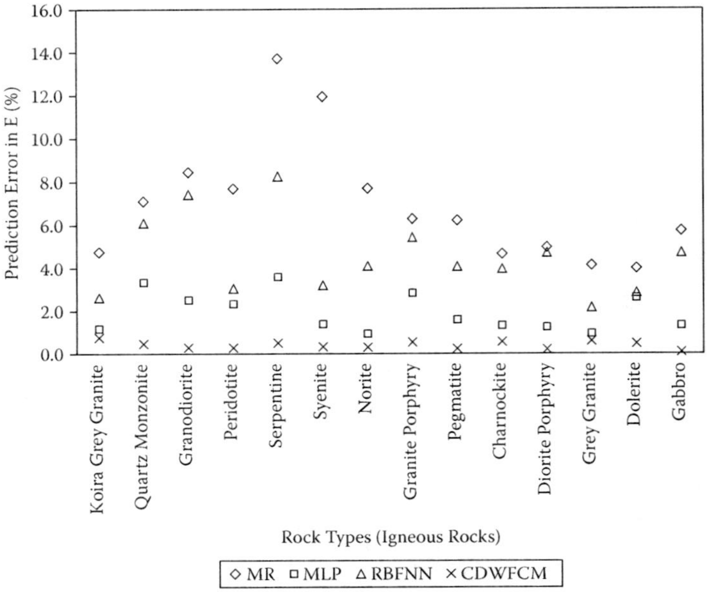

FIGURE 8.15
The error between the predicted values in Young's modulus (%) using the constructed models and the experimental values.

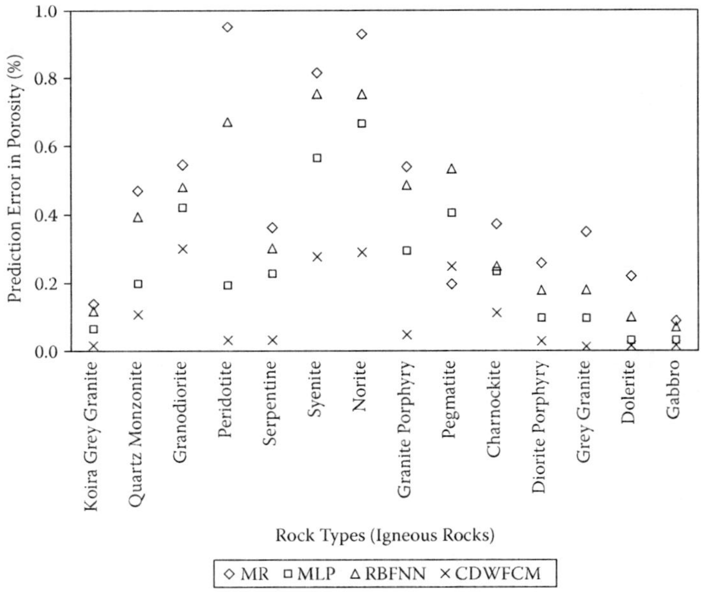

FIGURE 8.16
The error between the predicted values in porosity (%) using the constructed models and the experimental values.

deviation interval of the predicted values from CDWFCM with fixed width model is smaller than the deviation interval of multiple regression models, MLP and RBFNN with random selection of centers.

References

Anderson, J.A. (1995). *An Introduction to Neural Networks*. Cambridge, MA, MIT Press.

Hadjahmadi, A.H., Homayounpour, M.M., and Ahadi, S.M. (2008). "Robust weighted fuzzy C-means clustering," *Proc., IEEE Int. Conf. on Fuzzy Systems (FUZZ2008)*, 305–311.

Haykin, S. (1999). *Neural Networks—A Comprehensive Foundation*. Prentice Hall, Upper Saddle River, New Jersey.

Lowe, D., Webb, A.R. (1991), Time series prediction by adaptive networks: a dynamical systems perspective, Radar and Signal Processing, IEEE Proceedings F, Vol. 138, No. 1, pp. 17–24.

Magali, R.G.M., and Paul E.M.A. (2003). "A comprehensive review for industrial applicability of artificial neural networks." *IEEE Trans. Ind. Electron.*, 50(3), 585–601.

Neural Network Toolbox ™ 6 Mathworks—R2008.

9

Case Study—Sound Level versus Rock Properties

In the present work, an attempt was also made to experimentally determine the uniaxial compressive strength (UCS) in the field during drilling blast holes. The Medapalli Open Cast Project (MOCP), belonging to M/S Singareni Colliery Company Limited, situated in the state of Andhra Pradesh, India, was used for the field investigations. It is situated on the northwest end of Ramagundam Coal Field and is on the southern bank of the river Godavari.

9.1 Lithology of the Mine/Area

The rock stratum at the MOCP consists primarily of sandstone, carbonaceous shale, sandy shale, coal, shale, shaly coal, carbonaceous sandstone, and carbonaceous clay. There were a total of five coal seams in that area. Out of these five seams, four coal seams from the top had already been extracted. Bore-hole data near the investigation area are shown in Figure 9.1, which were obtained from the geology section of the mine. The lithological details from the fourth to fifth seam are also indicated in Figure 9.1 (right-hand side of the figure), with the depth of each rock formation. Between the fourth and fifth seam, the strata are classified into upper roof (3.0–6.0 m above the top of the coal seam, i.e., top of fifth seam), immediate roof (0.0–3.0 m above the top of the coal seam), immediate floor (0.0–3.0 m below the base of the coal seam), main floor (3.0–6.0 m below the base of the coal seam), and interburden (bounding strata not classified as roof or floor).

9.2 Drilling Machine and Sound Measuring Instrument

A rotary drill machine was used for drilling blast holes in the mine. The drill bit diameter was 150.0 mm with tungsten carbide button bits. Air was used as the flushing fluid. Compressed air was used as the feed mechanism with a sump pressure of 1.275 MPa and a line pressure of 1.373 MPa.

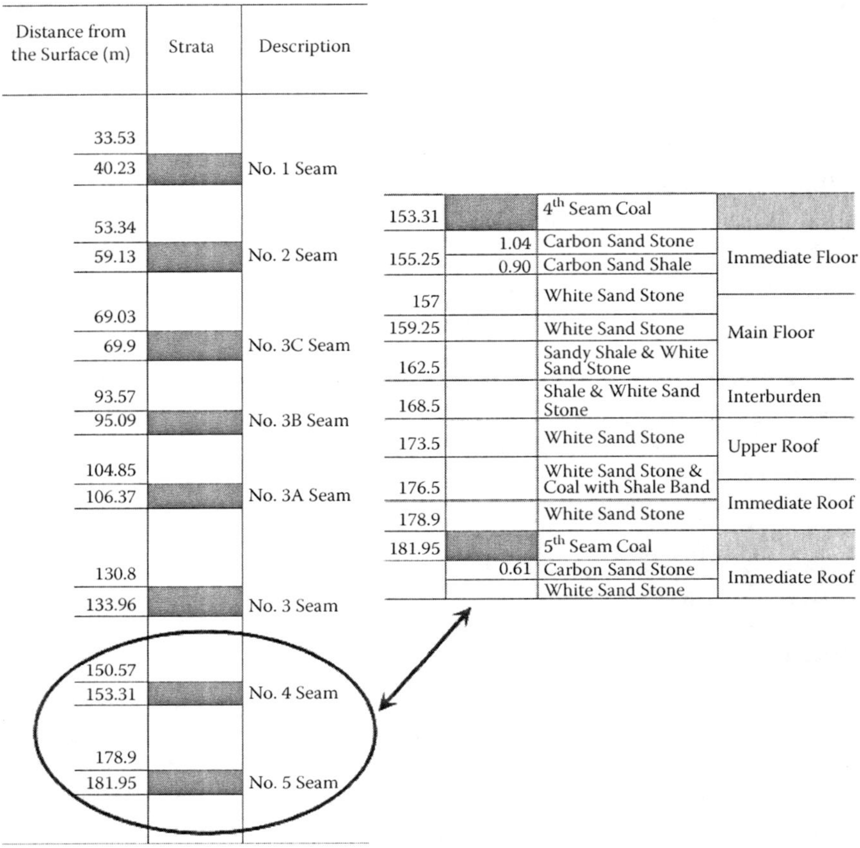

FIGURE 9.1
Lithology of the area (bore hole data).

The equivalent sound level measuring instrument was dosimeter for field investigations (which has been described in earlier chapters).

9.3 Sound Measurement

Field investigation of the sound levels produced during drilling was carried out on the rotary drill machine using dosimeter. All the measurements were carried out while drilling blast holes. During field investigation, bit type and diameter, blast hole length, weight on bit, compressed air pressure, net drilling time and rpm of the drill bit were recorded. The penetration rate (m/min) was calculated from the drilled hole length (meters) and the net drilling time (minutes). Blast holes were drilled between the fourth and fifth seams at each classified strata (Figure 9.1). Depending on the blast design, the blast hole length was limited to 6.0 m, whereas at other places it was only 3.0 m. For a 3.0 m long blast hole length, the weight on the bit was 12.0 kg, whereas for the 6.0 m long holes, the weight on the bit was 8.0 kg.

The exploratory bore-hole data were collected from the geology section of the mine. The UCS, density, tensile strength, Young's modulus, and impact strength of various strata were collected from the exploratory bore-hole data near the blast hole drilling as given in Table 9.1. For all sound measurements, the dosimeter was handheld at a height of 1.0 m from the ground level and at a distance of 1.5 m and 2.5 m from the blast hole (Figure 9.2). Sound levels were recorded for 16 different drill holes. At each second, the equivalent continuous A-weighted sound levels were recorded by the dosimeter. The data recorded during field measurements using the dosimeter were downloaded to the computer for analysis.

Some critical observations, such as color change of flushing dust and the exact time during color change, were also recorded. For the same drill diameter and type, penetration rate, and weight on bit, the sound levels were measured for various drilled holes, consisting of strata of different compressive strengths.

9.4 Sound Level Vis-à-Vis Rock Properties

In this work, using dosimeter, equivalent sound pressure Level (L_{eq}) was measured for each second. Drill bit penetration rate in m/sec was calculated. The time taken to drill a 3.0 m deep hole was noted down. Then L_{eq} versus drill hole depth was plotted and is as shown in Figures 9.3 and 9.4.

Investigation with 8.0 kg weight on bit during drilling was also carried out on blast hole #1, having white sandstone with compressive strength of 36.49 MPa and blast hole #12, containing sandy shale and white sandstone with compressive strength of 28.35 MPa. From the Figure 9.3, it is observed that for the first 45.0 cm depth of drilling, the difference in sound level for blast hole #1 and blast hole #12 is as much as 6.7 dB. By neglecting the first 45.0 cm depth, it is observed that for increase in compressive strength by 8.14 MPa (UCS of blast hole #1 and blast hole #12), L_{eq} level increases up to 4.0 dB.

Figure 9.4 shows results of investigation with 12.0 kg weight on bit during drilling. In this case, blast hole #2 was shale with white sandstone of compressive strength 30.61 MPa and blast hole #14 was white sandstone of compressive strength 37.08 MPa, whereas blast hole #6 was white sandstone and coal with shale band of compressive strength 28.84 MPa.

It is observed that for the first 45.0 cm depth of drilling the increase in sound level for blast hole #2 compared to that of blast hole #6 is as much as 2.2 dB. Similarly, the increase in sound level for blast hole #14 compared to that of blast hole #2 is as much as 4.9 dB. In addition, the sound level of blast

TABLE 9.1

Exploratory Borehole Data Near the Blast Hole Drilling

Drill Hole No.	Strata Location	Distance from Surface in Meters	Formation	Drill Bit Diameter (mm)	Weight on Bit (kg)	Drill Rod RPM	Observation Distance (m)	Penetration Rate (m/min)	Density (g/cc)	Tensile Strength (MPa)	Compressive Strength (MPa)	Young's Modulus (GPa)	Impact Strength Number
1	Immediate floor of 4th seam	155.25 to 159.25	White sandstone	150	8	85	1.5	0.82	2.28	2.62	36.49	5.10	52.02
2	Interburden between 4th and 5th seam	162.5 to 168.5	Shale and white sandstone	150	12	73	1.5	1.00	2.24	2.36	30.61	3.83	49.42
3	Interburden between 4th and 5th seam	162.5 to 168.5	Shale and white sandstone	150	12	73	2.5	1.00	2.24	2.36	30.61	3.83	49.42
4	Upper roof of 5th seam	173.5 to 176.5	White sandstone and coal with shale band	150	12	73	4.5	1.00	2.22	1.98	28.84	3.01	47.80
5	Upper roof of 5th seam	173.5 to 176.5	White sandstone and coal with shale band	150	12	73	3.5	1.00	2.22	1.98	28.84	3.01	47.80
6	Upper roof of 5th seam	173.5 to 176.5	White sandstone and coal with shale band	150	12	73	1.5	1.00	2.22	1.98	28.84	3.01	47.80

7	Upper roof of 5th seam	173.5 to 176.5	White sandstone and coal with shale band	150	12	73	2.5	1.00	2.22	1.98	28.84	3.01	47.80
8	Main floor after 4th seam	157 to 162	Sandy shale and white sandstone	150	8	85	2.5	0.82	2.21	1.81	28.35	2.75	47.32
9	Main floor after 4th seam	157 to 162	Sandy shale and white sandstone	150	8	85	3.5	0.82	2.21	1.81	28.35	2.75	47.32
10	Main floor after 4th seam	157 to 162	Sandy shale and white sandstone	150	8	85	4.5	0.82	2.21	1.81	28.35	2.75	47.32
11	Main floor after 4th seam	157 to 162	Sandy shale and white sandstone	150	8	85	5.5	0.82	2.21	1.81	28.35	2.75	47.32
12	Main floor after 4th seam	157 to 162	Sandy shale and white sandstone	150	8	85	1.5	0.82	2.21	1.81	28.35	2.75	47.32
13	Immediate roof of 5th seam	175.9 to 178.9	White sandstone	150	12	73	2.5	1.00	2.29	3.14	37.08	6.18	52.36
14	Immediate roof of 5th seam	175.9 to 178.9	White sandstone	150	12	73	1.5	1.00	2.29	3.14	37.08	6.18	52.36
15	Immediate roof of 5th seam	175.9 to 178.9	White sandstone	150	12	73	4.5	1.00	2.29	3.14	37.08	6.18	52.36
16	Immediate roof of 5th seam	175.9 to 178.9	White sandstone	150	12	73	3.5	1.00	2.29	3.14	37.08	6.18	52.36

FIGURE 9.2
Sound measurement during drilling.

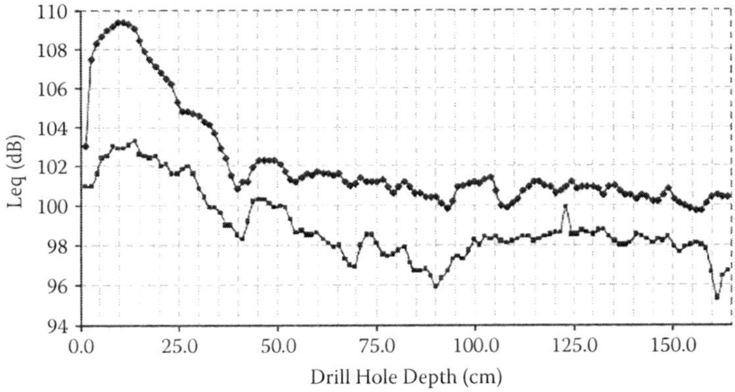

FIGURE 9.3
Comparison of L_{eq} vs. drill hole depth with 8.0 kg weight on drill bit: ◆ blast hole—1 (UCS 36.49 MPa); ■ blast hole—12 (UCS 28.35 MPa).

hole #14 is up to 6.7 dB higher than that of blast hole #6. By neglecting the first 45.0 cm depth, it is observed that for an increase in compressive strength by 1.77 MPa (UCS of blast hole #2 and blast hole #6), L_{eq} level increases up to 2.8 dB. For increase in compressive strength by 8.24 MPa (UCS of blast hole #14 and blast hole #6) L_{eq} level increases up to 8.0 dB. Similarly, for increase in compressive strength by 6.47 MPa (UCS of blast hole #14 and blast hole #2), L_{eq} level increases up to 7.1 dB.

This clearly indicates that as the compressive strength increases, the L_{eq} level produced during drilling also increases. However, this increase in L_{eq}

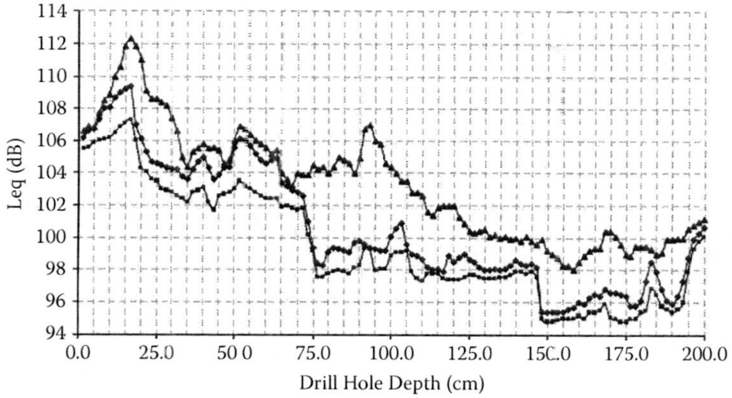

FIGURE 9.4
Comparison of L_{eq} vs. drill hole depth with 12.0 kg weight on drill bit: ♦ blast hole—2 (UCS 30.61 MPa); ■ blast hole—6 (UCS 28.84 MPa); ▲ blast hole—14 (UCS 37.08 MPa).

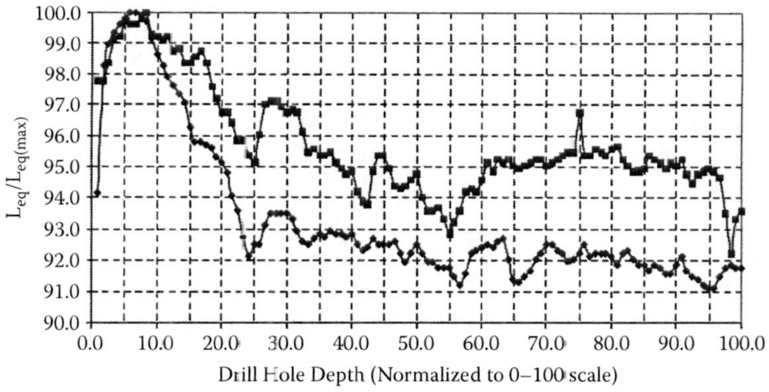

FIGURE 9.5
Variation of $L_{eq}/L_{eq(max)}$ with normalized values of drill hole depth for 8.0 kg weight on drill bit: ♦ blast hole—1 (UCS 36.49 MPa); ■ blast hole—12 (UCS 28.35 MPa).

level also depends on the weight on the bit, which is indirectly related to the compressor pressure used.

It is also observed that between depths of 75.0 cm to 125.0 cm and 150.0 cm to 175.0 cm the L_{eq} levels measured at blast hole #6 and blast hole #2 were somewhat similar, whereas blast hole #14 had an increase in L_{eq} value of up to 8.0 dB for depths between 75.0 cm and 125.0 cm and up to 5.3 dB for depths between 150.0 cm and 175.0 cm. This is because of the coal present in blast hole #6 and blast hole #2 between these depths, which was confirmed on observing the coal dust flushing out of the drill holes at these depths.

Figures 9.5 and 9.6 shows the plot of L_{eq}/L_{eq}(max) versus normalized values of drill hole depth in 0–100 scale. From both the figures, it can be observed

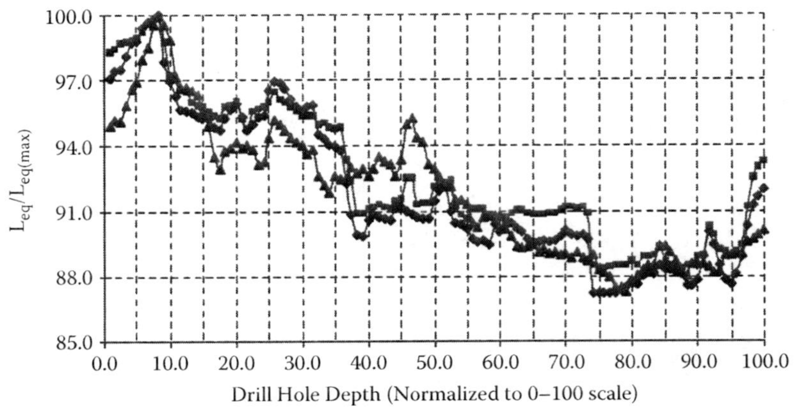

FIGURE 9.6
Variation of $L_{eq}/L_{eq(max)}$ with normalized values of drill hole depth for 12.0 kg weight on drill bit: ◆ blast hole—2 (UCS 30.61 MPa); ■ blast hole—6 (UCS 28.84 MPa); ▲ blast hole—14 (UCS 37.08 MPa).

TABLE 9.2

Comparison of A-Weighted Equivalent Sound Level (L_{eq}) and UCS for First 2 Minutes of Drilling

		L_{eq} (dB)									
UCS (MPa)	**Weight On a Bit (kg)**	**Blast Hole**	**1.5 m**	**Blast Hole**	**2.5 m**	**Blast Hole**	**3.5 m**	**Blast Hole**	**4.5 m**	**Blast Hole**	**5.5 m**
36.49	8.0	1	102.5	—	—	—	—	—	—	—	—
28.35	8.0	12	98.6	8	96.2	9	95.4	10	94.3	—	—
30.61	12.0	2	101.5	3	98.9	—	—	—	—	—	—
28.84	12.0	6	99.3	7	97.6	5	95.9	4	94.8	11	93.7
37.08	12.0	14	103.2	13	100.9	16	99.6	15	98.5	—	—

that the pattern of variation of equivalent sound level with the drill hole depth remains same irrespective of the compressive strength. However, in both the figures, the difference in equivalent sound level produced during drilling for various available UCS values does not remain constant throughout the drilling process. Hence, it can be concluded from the above figures that use of measured sound levels produced during drilling can only add to accuracy of prediction of UCS values when operating parameters of drilling are monitored.

Table 9.2 gives the equivalent A-weighted sound levels for blast holes of different compressive strengths at different measurement distances.

It was observed that as the measurement distance increases, the equivalent A-weighted sound level decreases. For example, 1.0 meter increase in distance from 1.5 m to 2.5 m, for UCS of 30.61 MPa (blast hole #2 and blast hole #3), the sound level decreased by 2.6 dB. Similar results were obtained at strata of different compressive strengths (blast holes #6, #7, #8, #12, #13, and #14). However, the available data were insufficient to develop the numerical models, which could otherwise be directly used for predicting the rock properties in the field.

10

Summary and Suggestions for Further Work

10.1 Summary

In the present work, drill bit parameters and equivalent sound level produced during drilling were used to predict various rock properties. Rock properties of various collected rocks were measured in the laboratory by ISRM-suggested methods. Rock drilling experiments were conducted using CNC machine with different bit diameter, speed, and penetration rates. For all the conditions, equivalent sound levels were recorded. The above experimental results were used to develop the prediction models using multiple regression and ANN techniques. The summary of the outcomes from the experimental and modeling work are as follows:

At constant bit speed, penetration rate, and for a particular bit diameter, equivalent sound level produced during drilling will increase with increase in rock density, P-wave velocity, hardness and strength (UCS, Young's modulus, and tensile strength), whereas it decrease with increase in porosity.

Multiple regression models presented in this book show statistically meaningful relationships with high-prediction performances, between rock properties with drilling parameters, and equivalent sound level produced during drilling.

In this work, MLP neural network was trained using *trainrp, traingdx, trainscg,* and *trainlm* algorithms, which are different means of implementing back-propagation algorithms. Their performances were compared in terms of VAF, RMSE, and MAPE values. *Trainlm* algorithms perform better than all other algorithms in the prediction of all the rock properties.

In this work, a comparison of a multiple regression model and MLP model using *trainlm* algorithm was done, which revealed that the MLP model will give better performance than multiple regression technique with lower MAPE and RMSE values and higher prediction accuracy (VAF value) for all the prediction variables.

RBF neural networks can be applied for predicting rock properties. Two center initialization strategies for the RBF units in the hidden layer was studied in this work, namely, random selection of centers and CDWFCM algorithms. For both center initialization approaches, networks were trained by using fixed and varying widths. The generalization capability of CDWFCM algorithm with fixed width strategy was found to be better than the other methods.

A comparison of MLP and RBF neural networks in terms of performance indices show that MLP with *trainlm* algorithm is better than RBFNN with random selection of centers. However, RBFNN with center initialized using CDWFCM and fixed width outperformed the prediction capability of the MLP model.

The performance comparison also showed that the neural network is a good approach for minimizing the uncertainties in the rock engineering projects. The ability to adapt or continue learning is another important advantage for rock parameters prediction because training data are limited and new cases are continuously encountered. For this reason, the use of neural network may provide new approaches and methodologies, and minimize the potential inconsistency of correlations.

Results from this work demonstrate that regression and neural network techniques yield similar results, and in general, neural network approach is efficient in predicting rock properties from the sound level produced during drilling. Furthermore, the developed models can be effectively used, with acceptable accuracy, at the preliminary stage of design to predict the rock properties in mining industries.

Overall, this work shows that use of measured sound level produced during drilling can add to accuracies of prediction of rock properties when operational parameters of drilling are measured and analyzed. However, this method may not be suitable for very hard rocks (UCS >130 MPa), since coolant is required to prevent burning of drill bit tip during drilling, which modifies the acoustic signature emitted from rock-drill interaction.

10.2 Suggestions for Further Work

In the present work, only drilling parameters were considered during the development of the prediction models. However, this work can be extended to develop the prediction models by considering the chemical compositions of the rocks like grain size, quartz content, etc.

The work presented in this book makes use of CNC machine available in the laboratory for drilling experiments. However, this is costly equipment.

Hence, to reduce the cost, it is recommended to develop a low-cost portable drilling machine with all features to control the penetration rate and variation in the drill bit speed.

In the present work, the models were developed to predict individual rock properties. However, analysis can be done to look into the interrelationship of different rock design parameters. Thus, various models can be developed only for dominant rock design parameters.

In this work, regression and ANN modeling techniques were used. However, other techniques like fuzzy modeling and wavelet techniques can be developed. Also, other analysis tools like fast Fourier transfer (FFT) can be used to transfer data to frequency domain and filter out some of the noise or to identify dominant frequencies that lead to identification of rock or rock properties.

In this work, only intact rocks were considered during drilling process and sound levels were recorded for the development of the prediction models. This work can be extended to rock mass drilling also.

Instead of carrying the rock samples to the laboratory, extensive field investigations can be done using the available drilling machines in the field itself, and prediction models can be developed using those data, so that developed models can be used directly to predict rock properties.

Appendix I

The observations made and results obtained during the analysis of metamorphic rock category are given in this section. The basic statistical evaluations of various rock properties of different metamorphic rocks are given in Table I.1

The multiple regression model to predict the uniaxial compressive strength of metamorphic rock types is

$$UCS = 543.14 + 1.776 \times A + 0.066 \times B - 12.05 \times D + 0.029 \times A^2 + 0.071 \times D^2 - 0.029 \times A \times D - 0.001 \times B \times D$$

The multiple regression model to predict Schmidt rebound number of the metamorphic rock types is

$$SRN = 63.7504 + 0.1472 \times A - 1.4976 \times D + 0.0109 \times A^2 + 0.0131 \times D^2 - 0.0065 \times A \times D$$

The multiple regression model to predict dry density of the metamorphic rock types is

$$\rho = 20.7364 + 0.0747 \times A - 0.3917 \times D + 0.0021 \times D^2 - 0.0008 \times A \times D$$

The multiple regression model to predict P-wave velocity of the metamorphic rock types is

$$V_p = -1.19265 - 0.02556 \times A + 0.22345 \times D - 0.00093 \times A^2 - 0.00154 \times D^2 + 0.00066 \times A \times D$$

The multiple regression model to predict tensile strength of the metamorphic rock types is

$$TS = 69.3158 + 0.2313 \times A + 0.0085 \times B + 0.3125 \times C - 1.5578 \times D + 0.0038 \times A^2 + 0.0093 \times D^2 - 0.0038 \times A \times D - 0.0001 \times B \times D - 0.0036 \times C \times D$$

The multiple regression model to predict Young's modulus of the metamorphic rock types is

$$E = 9.78834 - 0.14471 \times A + 0.78764 \times D + 0.00264 \times A^2 - 0.00192 \times D^2$$

TABLE I.1

Statistical Values of Metamorphic Rock Properties and Range of Equivalent Sound Level Values Obtained During Drilling

Sound Level		Rose Quartzite	Slate	Quartzite	Light Green Quartzite	Biotite Schist with Garnet	Pink Quartzite	Dark Black Quartzite	Green Quartzite	Eclogite	Gnesis Light Pink	Yellow Quartzite	Smoky Quartzite	Marble
UCS (MPa)	Min	92.2	78.2	110.1	66.7	26.2	61.2	90.1	83.2	81.8	108.1	72.3	56.3	85.3
	Max	98.1	84.7	118.1	74.2	32.2	64	97	87.4	84.1	111.9	76.4	61.3	91.2
	SD	2.318	2.541	3.163	3.433	2.379	1.030	2.675	1.849	0.972	1.541	1.623	1.903	2.276
	Mean	96.2	81.1	114.3	71	28.9	62.9	93.7	85.6	82.9	110.1	74.8	59.1	88.3
SRN	Min	66	58	67	55	27	50	66	59	59	67	54	47	60
	Max	68	60	70	57	30	52	68	62	61	69	57	52	64
	SD	0.707	1.000	1.225	0.707	1.225	0.707	0.707	1.414	0.707	0.707	1.225	1.871	1.414
	Mean	67	59	69	56	29	51	67	60	60	68	56	49	62
Density (gm/cc)	Min	2.9	2.69	3.81	2.555	2.275	2.58	2.789	2.68	2.67	2.81	3.135	2.46	2.87
	Max	2.98	2.795	3.98	2.91	2.44	2.78	3.016	2.99	2.885	2.98	3.4	2.71	3.235
	SD	0.030	0.045	0.068	0.140	0.069	0.094	0.095	0.129	0.081	0.067	0.110	0.091	0.133
	Mean	2.947	2.747	3.908	2.691	2.345	2.688	2.928	2.866	2.785	2.894	3.259	2.596	3.017
P-wave velocity (km/sec)	Min	5.319	4.294	6.809	3.803	3.763	4.013	4.806	4.493	4.377	6.118	4.238	3.892	4.554
	Max	5.691	4.411	6.952	4.321	3.861	4.101	4.901	4.632	4.453	6.189	4.399	3.982	4.678
	SD	0.137	0.049	0.063	0.223	0.036	0.035	0.041	0.058	0.031	0.027	0.059	0.037	0.052
	Mean	5.476	4.337	6.853	4.197	3.815	4.073	4.856	4.555	4.404	6.148	4.321	3.952	4.589

Tensile strength (MPa)	Min	10.98	9.89	14.05	8.10	2.94	6.89	11.35	10.00	9.37	13.00	9.00	6.70	10.00
	Max	13.12	10.10	14.19	9.00	3.60	8.60	11.87	11.69	10.90	14.06	9.50	7.50	11.91
	SD	1.120	0.090	0.050	0.390	0.260	0.810	0.200	0.670	0.580	0.500	0.210	0.300	0.740
	Mean	11.88	10.01	14.11	8.55	3.25	7.58	11.57	10.57	10.23	13.59	9.23	7.12	10.90
Young's modulus (GPa)	Min	74.90	73.94	75.30	71.95	62.00	70.86	74.56	74.50	74.19	74.88	72.55	70.40	74.61
	Max	75.30	74.50	75.50	72.90	62.78	71.50	75.40	75.03	75.00	75.40	73.50	70.90	75.30
	SD	0.161	0.204	0.075	0.359	0.277	0.270	0.313	0.218	0.377	0.202	0.398	0.179	0.283
	Mean	75.1	74.2	75.4	72.5	62.4	71.2	75.0	74.8	74.5	75.2	73.2	70.7	75.0
Porosity (%)	Min	0.2012	0.3890	0.1430	0.7500	2.9445	1.0000	0.2010	0.2690	0.3200	0.1800	0.5900	1.1100	0.2200
	Max	0.2188	0.3960	0.1580	0.7900	3.3210	1.1100	0.2400	0.2770	0.3590	0.1920	0.6240	1.2000	0.2500
	SD	0.0067	0.0029	0.0059	0.0174	0.1394	0.0540	0.0163	0.0034	0.0153	0.0050	0.0135	0.0368	0.0120
	Mean	0.2114	0.3925	0.1516	0.7753	3.1397	1.0609	0.2266	0.2735	0.3394	0.1879	0.6104	1.1714	0.237
L_{eq} (dB)	Min	116.9	113.9	119.4	109.4	85	106.2	116.4	114.9	114.1	118.9	112	105.1	116.2
	Max	121	119	123.2	114.3	88.3	110.6	120.2	119.5	119.6	122.1	117.1	109.9	120.1

The multiple regression model to predict percentage porosity of the metamorphic rock types is

$$n = 17.4127 + 0.0465 \times A - 0.2261 \times D - 0.0006 \times A^2 + 0.0007 \times D^2 - 0.0001 \times A \times D$$

Regression coefficient tables, ANOVA tables, and model summary tables for various rock properties of metamorphic rocks are given below from Table I.2 to Table I.8.

TABLE I.2A

Significance of Regression Coefficients for the Estimation of Uniaxial Compressive Strength (Metamorphic Rocks)

Model Terms for UCS	Parameter Estimate (Coefficients)	t-Value	p-Value
Constant	543.142	24.223	0.000
A	1.776	4.112	0.000
B	0.066	1.812	0.041
D	−12.050	−28.837	0.000
A^2	0.029	3.247	0.001
D^2	0.071	34.939	0.000
AD	−0.029	−8.576	0.000
BD	−0.001	−2.360	0.010

TABLE I.2B

Analysis of Variance (ANOVA) for the Selected Quadratic Model for the Estimation of UCS (Metamorphic Rocks)

Source of Variations	Degree of Freedom	Sum of Squares	Mean Squares	F-Value	p-Value
Model	7	287172.50	41024.60	2221.92	0.000
Linear	3	15994.50	5331.50	288.76	0.000
Square	2	22603.40	11301.70	612.11	0.000
Interaction	2	1455.50	727.80	39.42	0.000
Residual Error	568	10487.30	18.50	—	—
Total	575	297660	—	—	—

TABLE I.2C

Model Summary for Dependent Variable (UCS—Metamorphic Rocks)

R^2	Predicted R^2	Adjusted R^2	Standard Error
0.8648	0.8640	0.8643	4.29693

TABLE I.3A

Significance of Regression Coefficients for the Estimation of Schmidt Rebound Number (Metamorphic Rocks)

Model Terms for SRN	Parameter Estimate (Coefficients)	*t*-Value	*p*-Value
Constant	63.7504	6.558	0.000
A	0.1472	0.744	0.047
D	−1.4976	−7.895	0.000
A^2	0.0109	2.654	0.008
D^2	0.0131	14.069	0.000
AD	−0.0065	−4.142	0.000

TABLE I.3B

Analysis of Variance (ANOVA) for the Selected Quadratic Model for the Estimation of Schmidt Rebound Number (Metamorphic Rocks)

Source of Variations	Degree of Freedom	Sum of Squares	Mean Squares	*F*-Value	*p*-Value
Model	5	73581.88	14716.377	3798.48	0.000
Linear	2	241.87	120.934	31.21	0.000
Square	2	783.84	391.920	101.16	0.000
Interaction	1	66.48	66.478	17.16	0.000
Residual Error	570	2208.34	3.874	—	—
Total	575	75790.2	—	—	—

TABLE I.3C

Model Summary for Dependent Variable (SRN—Metamorphic Rocks)

R^2	Predicted R^2	Adjusted R^2	Standard Error
0.8711	0.8705	0.8708	1.96832

TABLE I.4A

Significance of Regression Coefficients in the Estimation of Dry Density (Metamorphic Rock)

Model Terms for Dry Density	Parameter Estimate (Coefficients)	*t*-Value	*p*-Value
Constant	20.7364	17.127	0.000
A	0.0747	3.404	0.001
D	−0.3917	−16.537	0.000
D^2	0.0021	17.804	0.000
AD	−0.0008	−4.065	0.000

TABLE I.4B

Analysis of Variance (ANOVA) for the Selected Quadratic Model for the Estimation of Dry Density (Metamorphic Rock)

Source of Variations	Degree of Freedom	Sum of Squares	Mean Squares	*F*-Value	*p*-Value
Model	4	58.1818	14.5455	240.62	0.000
Linear	2	16.8117	8.4058	139.05	0.000
Square	1	19.1621	19.1621	316.99	0.000
Interaction	1	0.9988	0.9988	16.52	0.000
Residual error	571	34.5170	0.0605	—	—
Total	575	92.6989	—	—	—

TABLE I.4C

Model Summary for Dependent Variable (Dry Density—Metamorphic Rock)

R^2	Predicted R^2	Adjusted R^2	Standard Error
0.7257	0.7239	0.7250	0.245866

TABLE I.5A

Significance of Regression Coefficients for the Estimation of P-Wave Velocity (Metamorphic Rock)

Model Terms for P-Wave Velocity	Parameter Estimate (Coefficients)	*t*-Value	*p*-Value
Constant	−1.19265	−2.251	0.025
A	−0.02556	−2.371	0.018
D	0.22345	21.612	0.000
A^2	−0.00093	−4.127	0.000
D^2	−0.00154	−30.324	0.000
AD	0.00066	7.783	0.000

TABLE I.5B

Analysis of Variance (ANOVA) for the Selected Quadratic Model for the Estimation of P-Wave Velocity—(Metamorphic Rock)

Source of Variations	Degree of Freedom	Sum of Squares	Mean Squares	*F*-Value	*p*-Value
Model	5	436.5571	87.311141	7586.88	0.000
Linear	2	5.3920	2.69598	234.27	0.000
Square	2	10.6804	5.34019	464.03	0.000
Interaction	1	0.6971	0.69710	60.57	0.000
Residual Error	570	6.5597	0.01151	—	—
Total	575	443.117	—	—	—

TABLE I.5C

Model Summary for Dependent Variable (P-Wave Velocity—Metamorphic Rock)

R^2	Predicted R^2	Adjusted R^2	Standard Error
0.8853	0.8848	0.8851	0.107276

TABLE I.6A

Significance of Regression Coefficients for the Estimation of Tensile Strength (Metamorphic Rock)

Model Terms for Tensile Strength	Parameter Estimate (Coefficients)	t-Value	p-Value
Constant	69.3158	26.135	0.000
A	0.2313	4.668	0.000
B	0.0085	2.045	0.041
C	0.3125	1.502	0.034
D	–1.5578	–32.523	0.000
A^2	0.0038	3.657	0.000
D^2	0.0093	39.504	0.000
AD	–0.0038	–9.698	0.000
BD	–0.0001	–2.663	0.008
CD	–0.0036	–1.980	0.048

TABLE I.6B

Analysis of Variance (ANOVA) for the Selected Quadratic Model for the Estimation of Tensile Strength (Metamorphic Rock)

Source of Variations	Degree of Freedom	Sum of Squares	Mean Squares	F-Value	p-Value
Model	9	4742.541	526.949	2168.27	0.000
Linear	4	269.629	67.407	277.37	0.000
Square	2	380.311	190.155	782.45	0.000
Interaction	3	25.380	8.460	34.81	0.000
Residual error	566	137.553	0.243	—	—
Total	575	4880.09	—	—	—

TABLE I.6C

Model Summary for Dependent Variable (Tensile Strength—Metamorphic Rock)

R^2	Predicted R^2	Adjusted R^2	Standard Error
0.9116	0.9104	0.9110	0.492978

TABLE I.7A

Significance of Regression Coefficients for the Estimation of Young's Modulus (Metamorphic Rock)

Model Terms for Young's Modulus	Parameter Estimate (Coefficients)	*t*-Value	*p*-Value
Constant	9.78834	7.142	0.000
A	–0.144471	–9.507	0.000
D	0.78764	29.590	0.000
A^2	0.00264	4.567	0.000
D^2	–0.00192	–14.949	0.000

TABLE I.7B

Analysis of Variance (ANOVA) for the Selected Quadratic Model for the Estimation of Young's Modulus (Metamorphic Rock)

Source of Variations	Degree of Freedom	Sum of Squares	Mean Squares	*F*-Value	*p*-Value
Model	4	8723.6133	2180.9033	28300.13	0.000
Linear	2	77.0965	38.5482	500.21	0.000
Square	2	19.0587	9.5294	123.66	0.000
Residual Error	571	44.0032	0.0771	—	—
Total	575	8767.62	—	—	—

TABLE I.7C

Model Summary for Dependent Variable (Young's Modulus—Metamorphic Rock)

R^2	Predicted R^2	Adjusted R^2	Standard Error
0.9053	0.9045	0.9049	0.277603

TABLE I.8A

Significance of Regression Coefficients for the Estimation of % Porosity (Metamorphic Rock)

Model Terms for % Porosity	Parameter Estimate (Coefficients)	*t*-Value	*p*-Value
Constant	17.4127	60.114	0.000
A	0.0465	7.889	0.000
D	–0.2261	–39.997	0.000
A^2	–0.0006	–4.507	0.000
D^2	0.0007	24.078	0.000
AD	–0.0001	–2.967	0.003

TABLE I.8B

Analysis of Variance (ANOVA) for the Selected Quadratic Model for the Estimation of % Porosity (Metamorphic Rock)

Source of Variations	Degree of Freedom	Sum of Squares	Mean Squares	*F*-Value	*p*-Value
Model	5	462.03735	92.40747	26861.94	0.000
Linear	2	5.61632	2.80816	816.30	0.000
Square	2	2.09720	1.04860	304.82	0.000
Interaction	1	0.03027	0.03027	8.80	0.003
Residual Error	570	1.96085	0.00344	—	—
Total	575	463.998	—	—	—

TABLE I.8C

Model Summary for Dependent Variable (% Porosity—Metamorphic Rock)

R^2	Predicted R^2	Adjusted R^2	Standard Error
0.9185	0.9180	0.9183	0.0586523

Appendix II

The observations made and results obtained during the analysis of sedimentary rock category are given in this section. The basic statistical evaluations of various rock properties, of different sedimentary rocks are given in Table II.1.

The multiple regression model to predict uniaxial compressive strength of the sedimentary rock types is

$$\begin{aligned} UCS = {} & 71.8640 + 0.2922 \times A + 0.0016 \times B + 0.0171 \times C - 2.3069 \times D \\ & + 0.0080 \times A^2 + 0.0217 \times D^2 + 0.0005 \times A \times B + 0.0167 \times A \times C \\ & - 0.0106 \times A \times D - 0.0002 \times B \times D - 0.0072 \times C \times D \end{aligned}$$

The multiple regression model to predict Schmidt rebound number of the sedimentary rock types is

$$\begin{aligned} \mathrm{SRN} = {} & -12.2623 - 0.067 \times \mathrm{A} - 0.0101 \times \mathrm{B} - 0.3875 \times \mathrm{C} + 0.1881 \times \mathrm{D} \\ & + 0.0042 \times D^2 + 0.0003 \times \mathrm{A} \times \mathrm{B} + 0.0083 \times \mathrm{A} \times \mathrm{C} - 0.0024 \times \mathrm{A} \times \mathrm{D} \end{aligned}$$

The multiple regression model to predict dry density of the sedimentary rock types is

$$\begin{aligned} \rho = {} & 5.54466 + 0.0179 \times A + 0.0006 \times B + 0.02485 \times C - 0.09541 \times D \\ & + 0.00024 \times A^2 + 0.00066 \times D^2 - 0.00031 \times A \times D - 0.00001 \times B \times D \\ & - 0.00032 \times C \times D \end{aligned}$$

The multiple regression model to predict P-wave velocity of the sedimentary rock types is

$$V_p = -0.999038 - 0.033844 \times \mathrm{A} + 0.14272 \times \mathrm{D} - 0.000869 \times D^2 + 0.000397 \times \mathrm{A} \times \mathrm{D}$$

The multiple regression model to predict tensile strength of the sedimentary rock types is

$$\begin{aligned} TS = {} & 14.7292 + 0.0597 \times \mathrm{A} + 0.0013 \times \mathrm{B} + 0.0487 \times \mathrm{C} - 0.4146 \times \mathrm{D} \\ & + 0.0012 \times A^2 + 0.0034 \times D^2 + 0.0001 \times \mathrm{A} \times \mathrm{B} + 0.0022 \times \mathrm{A} \times \mathrm{C} \\ & - 0.0016 \times \mathrm{A} \times \mathrm{D} - 0.0014 \times \mathrm{C} \times \mathrm{D} \end{aligned}$$

The multiple regression model to predict Young's modulus of the sedimentary rock types is

$$E = -14.8189 - 0.2453 \times \mathrm{A} + 1.203 \times \mathrm{D} - 0.0037 \times D^2 + 0.0014 \times \mathrm{A} \times \mathrm{D}$$

TABLE II.1

Statistical Values of Sedimentary Rock Properties and Range of Equivalent Sound Level Values Obtained during Drilling

		Sandstone	Ironstone	Shell Limestone	Marl	Shale	Limestone	Chalk
UCS (MPa)	Min	61.2	82.9	16.8	57.3	15	71.2	21
	Max	62.8	83.6	18	59.4	15.6	72.3	21.8
	SD	0.632	0.274	0.485	0.872	0.255	0.442	0.374
	Mean	62.2	83.2	17.2	58.3	15.2	71.8	21.3
SRN	Min	50	59	23	46	20	54	25
	Max	52	61	23	49	23	57	27
	SD	0.707	0.707	0.000	1.225	1.225	1.225	0.707
	Mean	51	60	23	48	22	56	26
Density (gm/cc)	Min	2.6013	3.0267	2.0632	2.5831	1.8163	2.6700	2.2248
	Max	2.8192	3.2271	2.2026	2.6502	2.2822	2.8023	2.3127
	SD	0.0859	0.0818	0.0597	0.0278	0.1756	0.0591	0.0326
	Mean	2.6841	3.0962	2.1431	2.6114	1.9997	2.7445	2.2581
P-wave velocity (km/sec)	Min	4.5815	4.8644	4.623	3.9216	4.498	3.875	4.2027
	Max	4.7185	5.111	4.7612	4.3572	4.65	4.1309	4.3176
	SD	0.054	0.1056	0.0688	0.1645	0.0618	0.1001	0.0427
	Mean	4.642	4.9491	4.6868	4.1666	4.5477	3.9724	4.2592

Tensile strength (MPa)	Min	7.4124	10.0345	2.1896	6.9242	1.9360	8.8112	2.6987
	Max	7.5316	10.5046	2.2304	7.0911	1.9626	8.9101	2.7960
	SD	0.0482	0.1735	0.0172	0.0684	0.0098	0.0361	0.0395
	Mean	7.4940	10.2716	2.2051	7.0241	1.9487	8.8642	2.7308
Young's modulus (GPa)	Min	68.8742	72.5430	56.9834	69.5944	53.0670	71.9954	61.1331
	Max	73.4112	79.0252	57.1749	71.5479	53.9976	73.6398	62.5552
	SD	1.6081	2.5779	0.0778	0.7568	0.4671	0.6860	0.6409
	Mean	71.0791	74.7816	57.0637	70.5537	53.6032	72.6360	61.9314
Porosity (%)	Min	1.0098	0.2548	4.4957	1.1158	5.3387	0.7218	3.2042
	Max	1.1249	0.2992	4.5921	1.2946	5.7459	0.7709	3.2891
	SD	0.0443	0.0159	0.0351	0.0663	0.1475	0.0191	0.0350
	Mean	1.0815	0.2793	4.5511	1.1987	5.5394	0.7392	3.2579
L_{eq} (dB)	Min	105.8	114.2	76.5	104.3	75.5	109.9	80.9
	Max	110.2	119.5	81.6	108.9	80.9	114.9	84.3

The multiple regression model to predict percentage porosity of the sedimentary rock types is

$$n = 25.7588 + 0.0777 \times A - 0.3721 \times D + 0.0013 \times D^2 - 0.0005 \times A \times D$$

Regression coefficient tables, ANOVA tables and model summary tables for various rock properties of sedimentary rocks are given from Table II.2 to Table II.8.

TABLE II.2A

Significance of Regression Coefficients for the Estimation of Uniaxial Compressive Strength (Sedimentary Rocks)

Model Terms for UCS	Parameter Estimate (Coefficients)	*t*-Value	*p*-Value
Constant	71.8640	24.508	0.000
A	0.2922	3.839	0.000
B	0.0016	2.345	0.030
C	0.0171	2.075	0.040
D	–2.3069	–38.358	0.000
A^2	0.0080	4.049	0.000
D^2	0.0217	66.741	0.000
AB	0.0005	3.463	0.001
AC	0.0167	2.551	0.011
AD	–0.0106	–22.225	0.000
BD	–0.0002	–4.412	0.000
CD	–0.0072	–3.279	0.001

TABLE II.2B

Analysis of Variance (ANOVA) for the Selected Quadratic Model for the Estimation of UCS (Sedimentary Rocks)

Source of Variations	Degree of Freedom	Sum of Squares	Mean Squares	*F*-Value	*p*-Value
Model	11	228213.72	20746.70	42152.47	0.000
Linear	4	726.62	181.65	369.08	0.000
Square	2	2192.39	1096.20	2227.21	0.000
Interaction	5	258.24	51.65	104.94	0.000
Residual error	308	151.59	0.49	—	—
Total	319	228365	—	—	—

TABLE II.2C

Model Summary for Dependent Variable (UCS—Sedimentary Rocks)

R^2	Predicted R^2	Adjusted R^2	Standard Error
0.8993	0.8993	0.8993	0.701557

TABLE II.3A

Significance of Regression Coefficients for the Estimation of Schmidt Rebound Number (Sedimentary Rocks)

Model Terms for SRN	Parameter Estimate (Coefficients)	t-Value	p-Value
Constant	−12.2623	−6.182	0.000
A	−0.0670	−[illegible].597	0.011
B	−0.0101	−7.754	0.000
C	−0.3875	−5.961	0.000
D	0.1881	4.424	0.000
D^2	0.0042	18.323	0.000
AB	0.0003	2.763	0.006
AC	0.0083	1.798	0.043
AD	−0.0024	−7.004	0.000

TABLE II.3B

Analysis of Variance (ANOVA) for the Selected Quadratic Model for the Estimation of Schmidt Rebound Number (Sedimentary Rocks)

Source of Variations	Degree of Freedom	Sum of Squares	Mean Squares	F-Value	p-Value
Model	8	76390.2317	9548.779	38583.03	0.000
Linear	4	35.9678	8.992	36.33	0.000
Square	1	83.0927	83.092	335.75	0.000
Interaction	3	14.4963	4.8321	19.52	0.000
Residual Error	311	76.9683	0.2475	—	—
Total	319	76467.2	—	—	—

TABLE II.3C

Model Summary for Dependent Variable (SRN—Sedimentary Rocks)

R^2	Predicted R^2	Adjusted R^2	Standard Error
0.8691	0.8685	0.8689	0.497480

TABLE II.4A

Significance of Regression Coefficients for the Estimation of Dry Density (Sedimentary Rock)

Model Terms for Dry Density	Parameter Estimate (Coefficients)	*t*-Value	*p*-Value
Constant	5.54466	28.486	0.000
A	0.01790	3.977	0.000
B	0.00060	2.074	0.039
C	0.02485	1.716	0.047
D	−0.09541	−23.673	0.000
A^2	0.00024	1.842	0.036
D^2	0.00066	30.181	0.000
AD	−0.00031	−9.599	0.000
BD	−0.00001	−2.664	0.008
CD	−0.00032	−2.188	0.029

TABLE II.4B

Analysis of Variance (ANOVA) for the Selected Quadratic Model for the Estimation of Dry Density (Sedimentary Rock)

Source of Variations	Degree of Freedom	Sum of Squares	Mean Squares	*F*-Value	*p*-Value
Model	9	49.18265	5.46474	2472.23	0.000
Linear	4	1.27855	0.31964	144.60	0.000
Square	2	2.01348	1.00674	455.45	0.000
Interaction	3	0.22392	0.07464	33.73	0.000
Residual Error	310	0.68524	0.00221	—	—
Total	319	49.8679	—	—	—

TABLE II.4C

Model Summary for Dependent Variable (Dry Density—Sedimentary Rock)

R^2	Predicted R^2	Adjusted R^2	Standard Error
0.8555	0.8549	0.8552	0.0470154

TABLE II.5A

Significance of Regression Coefficients for the Estimation of P-Wave Velocity (Sedimentary Rock)

Model Terms for P-wave Velocity	Parameter Estimate (Coefficients)	*t*-Value	*p*-Value
Constant	–0.999038	–2.322	0.021
A	–0.033844	–4.596	0.000
D	0.142720	15.246	0.000
D^2	–0.000869	–17.243	0.000
AD	0.000397	5.335	0.000

TABLE II.5B

Analysis of Variance (ANOVA) for the Selected Quadratic Model for the Estimation of P-Wave Velocity (Sedimentary Rock)

Source of Variations	Degree of Freedom	Sum of Squares	Mean Squares	*F*-Value	*p*-Value
Model	4	32.9399	8.23496	686.04	0.000
Linear	2	2.8190	1.40951	117.42	0.000
Square	1	3.6126	3.6123	300.96	0.000
Interaction	1	0.3417	0.34163	28.47	0.000
Residual Error	315	3.7811	0.01200	—	—
Total	319	36.7210	—	—	—

TABLE II.5C

Model Summary for Dependent Variable (P-wave Velocity Sedimentary Rock)

R^2	Predicted R^2	Adjusted R^2	Standard Error
0.8973	0.8965	0.8970	0.109561

TABLE II.6A

Significance of Regression Coefficients for the Estimation of Tensile Strength (Sedimentary Rock)

Model Terms for Tensile Strength	Parameter Estimate (Coefficients)	t-Value	p-Value
Constant	14.7292	37.481	0.000
A	0.0597	5.851	0.000
B	0.0013	2.143	0.033
C	0.0487	2.600	0.041
D	–0.4146	–51.437	0.000
A^2	0.0012	4.709	0.000
D^2	0.0034	77.399	0.000
AB	0.0001	3.167	0.002
AC	0.0022	2.462	0.014
AD	–0.0016	–25.298	0.000
CD	–0.0014	–4.615	0.000

TABLE II.6B

Analysis of Variance (ANOVA) for the Selected Quadratic Model for the Estimation of Tensile Strength (Sedimentary Rock)

Source of Variations	Degree of Freedom	Sum of Squares	Mean Squares	F-Value	p-Value
Model	10	3332.8330	302.9848	34275.66	0.000
Linear	4	23.6270	5.9068	668.21	0.000
Square	2	52.9553	26.4777	2995.33	0.000
Interaction	4	6.0791	1.2158	137.54	0.000
Residual Error	308	2.7226	0.0088	—	—
Total	319	3335.56	—	—	—

TABLE II.6C

Model Summary for Dependent Variable (Tensile Strength—Sedimentary Rock)

R^2	Predicted R^2	Adjusted R^2	Standard Error
0.8892	0.8891	0.8891	0.0940194

TABLE II.7A

Significance of Regression Coefficients for the Estimation of Young's Modulus (Sedimentary Rock)

Model Terms for Young's Modulus	Parameter Estimate (Coefficients)	t-Value	p-Value
Constant	–14.8189	–3.725	0.000
A	–0.2453	–3.603	0.000
D	1.2030	13.901	0.000
D^2	–0.0037	–7.925	0.000
AD	0.0014	2.033	0.043

TABLE II.7B

Analysis of Variance (ANOVA) for the Selected Quadratic Model for the Estimation of Young's Modulus (Sedimentary Rock)

Source of Variations	Degree of Freedom	Sum of Squares	Mean Squares	F-Value	p-Value
Model	4	22427.994	5606.9985	5465.64	0.000
Linear	2	198.919	99.4594	96.95	0.000
Square	1	64.435	64.4351	62.81	0.000
Interaction	1	4.241	4.2415	4.13	0.043
Residual error	315	323.147	1.0259	—	—
Total	319	22751.1	—	—	—

TABLE II.7C

Model Summary for Dependent Variable (Young's Modulus—Sedimentary Rock)

R^2	Predicted R^2	Adjusted R^2	Standard Error
0.8661	0.8653	0.8656	1.01285

TABLE II.8A

Significance of Regression Coefficients for the Estimation of % Porosity (Sedimentary Rock)

Model Terms for % Porosity	Parameter Estimate (Coefficients)	t-Value	p-Value
Constant	25.7588	22.750	0.000
A	0.0777	4.008	0.000
D	–0.3721	–15.103	0.000
D^2	0.0013	9.907	0.000
AD	–0.0005	–2.671	0.008

TABLE II.8B

Analysis of Variance (ANOVA) for the Selected Quadratic Model for the Estimation of % Porosity (Sedimentary Rock)

Source of Variations	Degree of Freedom	Sum of Squares	Mean Squares	F-Value	p-Value
Model	4	1386.7642	346.69105	4168.94	0.000
Linear	2	19.0497	9.52486	114.54	0.000
Square	1	8.1623	8.16232	98.15	0.000
Interaction	1	0.5934	0.59336	7.14	0.008
Residual error	315	26.1956	0.08316	—	—
Total	319	1412.96	—	—	—

TABLE II.8C

Model Summary for Dependent Variable (% Porosity - Sedimentary Rock)

R^2	Predicted R^2	Adjusted R^2	Standard Error
0.8815	0.8809	0.8813	0.288376

Index

M

N

U

V

W

Y

Z